Prader-Willi Syndrome

and Other Chromosome 15q Deletion Disorders

NATO ASI Series

Advanced Science Institutes Series

A series presenting the results of activities sponsored by the NATO Science Committee, which aims at the dissemination of advanced scientific and technological knowledge, with a view to strengthening links between scientific communities.

The Series is published by an international board of publishers in conjunction with the NATO Scientific Affairs Division

A Life Sciences **B Physics**	Plenum Publishing Corporation London and New York
C Mathematical and Physical Sciences **D Behavioural and Social Sciences** **E Applied Sciences**	Kluwer Academic Publishers Dordrecht, Boston and London
F Computer and Systems Sciences **G Ecological Sciences** **H Cell Biology** **I Global Environmental Change**	Springer-Verlag Berlin Heidelberg New York London Paris Tokyo Hong Kong Barcelona Budapest

NATO-PCO DATABASE

The electronic index to the NATO ASI Series provides full bibliographical references (with keywords and/or abstracts) to more than 30000 contributions from international scientists published in all sections of the NATO ASI Series. Access to the NATO-PCO DATABASE compiled by the NATO Publication Coordination Office is possible in two ways:

- via online FILE 128 (NATO-PCO DATABASE) hosted by ESRIN, Via Galileo Galilei, I-00044 Frascati, Italy.
- via CD-ROM "NATO-PCO DATABASE" with user-friendly retrieval software in English, French and German (© WTV GmbH and DATAWARE Technologies Inc. 1989).

The CD-ROM can be ordered through any member of the Board of Publishers or through NATO-PCO, Overijse, Belgium.

Series H: Cell Biology Vol. 61

Prader-Willi Syndrome

and Other Chromosome 15q Deletion Disorders

Edited by

Suzanne B. Cassidy

University of Arizona
College of Medicine
Section of Genetics/Dysmorphology
Tucson, Arizona 85724
USA

Springer-Verlag
Berlin Heidelberg New York London Paris Tokyo
Hong Kong Barcelona Budapest
Published in cooperation with NATO Scientific Affairs Division

Proceedings of the NATO Advanced Research Workshop on Prader-Willi Syndrome and Other Chromosome 15q Deletion Disorders held at Noordwijkerhout (The Netherlands) from May 2–3, 1991

Additional material to this book can be downloaded from http://extra.springer.com.

ISBN-13:978-3-642-84285-6 e-ISBN-13:978-3-642-84283-2
DOI: 10.1007/978-3-642-84283-2

Library of Congress Cataloging-in-Publication Data
Prader-Willi Syndrome and other chromosome 156q deletion disorders / edited by Suzanne B. Cassidy.
(NATO ASI series. Series H, Cell biology; vol. 61 "Proceedings of the NATO Advanced workshop on Prader-Willi Syndrome and Other Chromosome 15q Deletion Disorders, held at Noordwijkerhout (The Netherlands), from May 2-3, 1991" – T. p. verso. "Published in cooperation with NATO Scientific Affairs Division." Includes bibliographical references.
ISBN-13:978-3-642-84285-6 (alk. paper)

1. Prader-Willi Syndrome-Congresses. 2. Angelman Syndrome-Congresses. 3. Human chromosome 15-Abnormalities-Congresses. I. Cassidy. Suzanne B. II. North Atlantic Treaty Organization. Scientific Affairs Division. III. NATO Advanced Research Workshop on Prader-Willi Syndrome and Other Chromosome 15q Deletion Disorders (1991 : Noordwijkerhout, The Netherlands) [DNLM: 1. Chromosome Abnormalities. 2. Chromosomes, Human, Pair 15. 3. Prader-Willi Syndrome-Congresses. QS 677 P896 1991] RJ520.P7P73 1992
DNLM/DLC

Softcover reprint of the hardcover 1st edition 1992

Typesetting: Camera-ready by authors
31/3145-5 4 3 2 1 0 – Printed on acid-free paper

TABLE OF CONTENTS

FOREWORD

Although Prader-Willi syndrome was first described 35 years ago, it was following detection of an interstitial chromosome 15q deletion in some affected patients ten years ago that it became a major focus of multidisciplinary scientific interest. This interest was compounded by the later determination that some patients with a clinically distinct disorder, Angelman syndrome, apparently also had the same chromosome 15q deletion. Subsequently, molecular genetic studies showed that some cytogenetically normal patients with both disorders have uniparental disomy, maternal in Prader-Willi syndrome and paternal in Angelman syndrome. Genetic imprinting has been implicated in this unusual phenomenon.

This Workshop was conceived to bring together clinical and basic scientists from around the world whose research was focused on unraveling this unique genetic situation and further delineating these two fascinating disorders. As this volume demonstrates, it was successful in reaching this goal. Laboratory and clinical scientists from 15 countries in four continents participated, and even more countries were represented among the professional and parent observers of its proceedings. Many participants had previously known each other in print only. As a consequence of the Workshop, conclusions could be drawn on several issues. International collaborative research efforts were established. And acquaintances were developed between people who investigate the genetics of these disorders from differing perspectives, resulting in enrichment of approach to answering the complex questions posed by these fascinating conditions. Plans were initiated for another such scientific workshop a few years hence.

This volume includes papers presented from the platform. Abstracts of these and the poster presentations have been published in the American Journal of Medical Genetics (Cassidy, 1991) along with a conference summary.

I wish to express extreme gratitude to Maryann Wagner for the enormous efforts involved in the typing and preparation of this volume, and to thank Henk Moezelaar and Jean Phillips-Martinsson for organizational assistance with the Workshop.

Suzanne B. Cassidy
Director of the NATO ARW

PREFACE

PRADER-LABHART-WILLI SYNDROME IN 1956 AND 1991

As one of the authors who first described this syndrome in 9 patients aged 5 to 23 years, I am happy to be alive to witness the increasing interest in this syndrome among pediatricians and geneticists and to see how it has become important as one of the first examples in man of the new genetic mechanisms of parental disomy and genomic imprinting.

It gives me great pleasure to have been able to welcome all of the participants and observers at this first international conference on the Prader-Willi Syndrome and Other Chromosome 15q Deletion Disorders. I wish to express my gratitude to Dr. Suzanne Cassidy for her initiative and organization of the preparation of the research workshop, to Mrs. Jean Phillips-Martinsson for the preparation of the conference for parents and professionals and to Mr. Henk Moezelaar for the local organization.

In our original paper (Prader et al, 1956) - it was in German and had only two pages - I was the junior author with primary interest in pediatric endocrinology and clinical genetics. The two senior authors were Heinrich Willi, a neonatologist and practicing pediatrician who lived from 1890 to 1971, and Alexis Labhart, a clinical internist and endocrinologist, who is now 75 years old. I would also like to mention Hans Zellweger, my clinical teacher and friend, who died last year at the age of 80 years and whom many of you have known. His enthusiasm for muscular disorders and genetics had a great influence on me.

In retrospect, our paper described most of the symptoms and clinical findings including the endocrine aspects and the absence of affected siblings. It was only later that we became aware of the importance of fetal history, of sticky saliva, of the fact that some patients are obese without being overweight, and of the diminished sensitivity to pain.

Concerning the etiology of the syndrome, we speculated among ourselves about a possible chromosomal disorder but rather favored dominant inheritance with each patient having a new mutation and being unable to reproduce. This scientific workshop showed how much more we know at present.

Concerning management, we have learned from the parents more than they from us, and there are many parents, educators and psychologists who have gained my admiration for their patience and skills in successfully managing the problems of overeating and abnormal behavior.

The most pressing of the unsolved problems is the pathogenesis of the insatiable hunger and its pharmacological modification. The answer will probably come from the field of neurotransmitters.

I am convinced that this Workshop will advance our scientific knowledge and stimulate further research about this syndrome as well as improving the practical management of these patients.

Andrea Prader, M.D.

Department of Pediatrics
University of Zurich
Kinderspital
Steinwiesstrasse 75
CH-8032 Zurich
Switzerland

Prader A, Labhart A, Willi H (1956) Ein syndrom von adipositas, kleinwuchs, kryptorchismus und oligophrenie nach myatonieartigem zustand im neugeborenenalter. Schweiz Med Wochenschr 86:1260

INTRODUCTION AND OVERVIEW OF PRADER-WILLI SYNDROME

Suzanne B. Cassidy, M.D.
Section of Genetics/Dysmorphology
Department of Pediatrics and
Steele Memorial Children's Research Center
University of Arizona College of Medicine
Tucson, Arizona, U.S.A.

In 1680, a painter to the Spanish court of King Charles II, Juan Carreno de Miranda, immortalized on canvas a six year old obese girl who was known for her "feeblemindedness" and "bad temper". She has almond-shaped palpebral fissures, a down-turned mouth, small hands, and central obesity, all characteristic of Prader-Willi syndrome. She hangs in the Prado Museum in Madrid, clad in one painting, nude in the other, a testimony to the fact that the condition we now know as Prader-Willi syndrome has existed for centuries. But it was not until 1956 that Drs. Prader, Labhart and Willi first described the pattern of abnormalities which includes low muscle tone, incomplete sexual development, obesity, and mental handicap (Prader et al, 1956) and which now bears their names.

Prader-Willi syndrome is a complex, multisystem disorder which can be difficult to diagnose and frustrating to manage. Unraveling its genetic basis has provided a puzzling and often unique challenge to scientists. This chapter contains a brief review of the phenotype, natural history and etiology of Prader-Willi syndrome (PWS). It is intended to provide a framework for the scientific papers contained in the subsequent chapters, as it provided a framework for the oral and poster presentations which were made at the workshop held in The Netherlands in early May, 1991. The areas of knowledge and expertise in the workshop audience was diverse--clinical scientists such as geneticists, endocrinologist,

NATO ASI Series, Vol. H 61
Prader-Willi Syndrome
Edited by S. B. Cassidy

developmentalists, psychiatrists and psychologists as well as laboratory scientists such as basic molecular geneticist and cytogeneticists. This diversity was a unique strength of the workshop. A uniformity of background understanding about Prader-Willi syndrome added to this strength.

The major features of PWS are listed in Table I. As in any syndrome, or pattern of anomalies, no one feature is essential or uniformly found in PWS, and all findings may not be present in any given individual.

Table I. Cardinal Features of Prader-Willi Syndrome.

- Infantile Hypotonia
- Hypogonadism (Genital Hypoplasia; Abnormal Puberty)
- Developmental Delay/Mental Deficiency
- Behavior Disorder/Characteristic Personality
- Obesity
- Dysmorphic Features
- Short Stature for Genetic Background

Angelman syndrome is a distinct disorder whose inclusion in this workshop is based upon its genetic relationship to PWS. Soon after an interstitial deletion of chromosome 15 long arm was determined to be present in a majority of people with PWS, a number of children with features not typical of PWS were found to have an apparently overlapping deletion. Many of these children were eventually determined to have Angelman syndrome. This workshop provided great advances in unraveling the genetic relationship between these two disorders. The volume of literature on Angelman syndrome is far less than that on PWS, in part because it was only described in 1965. A review of the features of Angelman syndrome follows that of PWS. A number of scientific presentations at the workshop added to knowledge about this condition.

PHENOTYPE AND NATURAL HISTORY OF PWS

One of the characteristics of PWS which complicates its

diagnosis is the marked difference in features and symptoms in affected infants compared with older children or adults. For this reason, a description of these features cannot easily be separated from the natural history, and so they will be described together and roughly chronologically. Numbers in parentheses following a named feature represent the percentage of patients who have that features, based upon literature review and personal experience (for example, Holm et al, 1981; Laurance et al, 1981; Bray et al, 1983; Cassidy et al, 1984; Butler et al, 1987; Butler, 1990; Greenswag, 1987).

PRENATAL FEATURES

Prior to birth, the presence of prenatal hypotonia in patients with PWS causes **decreased fetal movement** (80%-90%) and abnormal fetal position at delivery (20%-30%). Prematurity or postmaturity is common, and the need for assisted delivery or cesarean section is increased (20%-30%).

NEONATAL PERIOD AND INFANCY

Infantile hypotonia is a nearly universal finding, causing decreased movement, lethargy with decreased arousal, weak cry, and poor reflexes. The hypotonia is central in origin, so that investigation for neuromuscular disorders yields normal results, though muscle biopsy may show non-specific abnormality likely due to disuse atrophy (Zellweger, 1981). Poor suck and gag reflexes cause **feeding problems**, requiring special feeding techniques such a gavage feeding, spoon feeding, or special nipples (90%-100%). The feeding problems cause a period of "**failure to thrive**", which improves in a matter of weeks or months when the feeding problems resolve. **Motor developmental delay** (90%-100%) results from the hypotonia as well, with early milestones being reached in about twice the normal time (e.g., mean time to sitting is 12 months, to walking is 24 months). The hypotonia improves with age, over a period varying from a few to many months and occasionally up to two years, but is never completely normal.

Sex hormone secretion is subnormal prenatally (as well as postnatally), and results in **hypogonadism** which in infancy is manifested as **genital hypoplasia** (90%-100%). Males have a hypoplastic scrotum (thin, small, poorly rugated and pigmented) and **cryptorchidism** (70%-80%), which may be unilateral or bilateral. The penis may be small. Females have small labia minora and clitoris, though this may be difficult to detect. The small genitalia lasts throughout life if not treated. The cryptorchidism may resolve spontaneously even as late as adolescence (Uehling, 1980), or may require hormonal or surgical treatment if testicular descent is desired.

The characteristic **dysmorphic features** of PWS (Table II) may not be apparent at birth, and may evolve over time. However, abnormal features are almost always present in infancy (Aughton and Cassidy, 1990).

Table II. Dysmorphic Features of Prader-Willi Syndrome.

Narrow Bifrontal Diameter
Almond-shaped Palpebral Fissures
Upslanting Palpebral Fissures
Thin Down-turned Mouth
Abnormal Ear Lobe Shape
Small Hands and Feet by 10 Years
Straight Ulnar Border

EARLY CHILDHOOD (1-4 years)

While motor development continues to be slow, **delayed speech development** now becomes apparent (80%-100%). **Articulation abnormalities** are also common (80%-90%), and continue throughout life. **Hypopigmentation** (40%-60%) due to oculocutaneous albinoidism (Hittner et al, 1982) is present and results in fair hair, skin and eye color, and may contribute to **strabismus** (60%-70%) and **visual acuity abnormality** (30%-50%).

This period is usually most striking for the onset of **hyperphagia** (80%-90%), which results in **obesity** in most cases if not controlled. Weight distribution is central, with relative sparing of the distal extremities. Although the exact etiology of the excessive appetite and lack of satiety is unknown, it appears to be hypothalamic in origin. Several other factors contribute to the obesity, including decreased caloric requirement (Holm and Pipes, 1976) and inactivity. Both these features are likely related to decreased muscle mass and bulk due to hypotonia.

CHILDHOOD (4-12 years)

Cognitive abnormalities become most apparent as the child reaches school age (90%-100%). While IQ testing indicates that the average person with PWS is mildly retarded, the range is from severely retarded to not retarded, with about 40% having borderline retardation or low normal intelligence (Holm, 1981). Regardless of IQ, most children have multiple severe learning disabilities, and their academic performance is poor for their mental abilities.

The **behavioral disorder** (70%-100%) may begin as early as 2 years, but becomes most apparent as obesity begins. **Food-seeking behavior**, including hording or foraging for food, eating of inedibles, and stealing food or money to buy food, is almost universal and lasts throughout the life of the individual. Episodic temper tantrums, stubbornness, and controlling or manipulative behavior initially relate primarily to withholding of food, but later become generalized. These episodes punctuate a personality which at times is gentle and sweet and at other times can be perseverative and characterized as obsessive-compulsive (Whitman and Accardo, 1987). Poor social and peer interactions are common.

By age 10-12, the hands and feet have become measurably small (80%-90%). **Skin picking** (70%-80%), **decreased pain**

sensitivity (50%-60%), and **high vomiting threshold** (70%-80%) are common. **Seizures** may occur (5%-20%), and **scoliosis** is increased in frequency (40%-60%), occasionally requiring surgical intervention.

ADOLESCENCE AND ADULTHOOD

Obesity is the major cause of morbidity and mortality in PWS. The consequences of uncontrolled obesity are listed in Table III. **Diabetes** (5%-10%) is weight related, and though it is of the non-insulin dependent type, insulin is often required if diet is uncontrolled. Obesity can, however, be controlled in most cases by a combination of low calorie diet (usually less than 1200 kcal/day), exercise, and supervision, as well as avoidance of exposure to high calorie or extra food.

Table III. Consequences of Obesity in Prader-Willi Syndrome.

Cardiopulmonary Compromise
Diabetes Mellitus (Type II)
Somnolence
Poor Social Acceptance
Poor Self-Image

Abnormal pubertal development (80%-100%) results from the hypogonadotrophic hypogonadism which is due to hypothalamic insufficiency. Precocious adrenarche is not uncommon, but in general puberty is delayed, disordered and incomplete. In females, oligomenorrhea or amenorrhea are found (80%-100%). In males, poor beard and body hair growth, poor muscular development, lack of voice change and phallic and testicular enlargement are found. **Infertility** is uniform.

Short stature for the family (80%-100%) may be evident in childhood, but commonly results from lack of adolescent growth spurt (Holm, 1981; Bray et al, 1983). Although growth hormone stimulation testing may be normal, neurosecretory growth hormone deficiency has recently been demonstrated in many cases, and early studies show good response to growth hormone therapy (Lee et al, 1987). **Osteoporosis** may be a consequence of hypogonadism and inactivity.

Oral pathology is common, including **abnormal saliva,** which is viscous and ropy and decreased in volume. Frequent caries have been noted. **Malocclusion** is also common.

Psychological and emotional problems increase at this time (20%-40%), with true psychotic or schizophrenic episodes occurring at increased frequency but usually responding well to standard medications. **Sleep disturbances** have recently been described in PWS, including hypoventilation, sleep-onset REM, cataplexy, excessive daytime sleepiness, and occasional central or obstructive apnea.

ETIOLOGY OF PWS

In 1956 when PWS was initially described, it was proposed that some of the features reflected hypothalamic dysfunction (Prader et al, 1956). Evidence has supported this proposed mechanism, since many features of PWS are explainable on the basis of hypothalamic insufficiency. Indeed, individuals with traumatic or surgical hypothalamic lesions show many similar symptoms to PWS, including hyperphagia, hypogonadotrophic hypogonadism, growth hormone deficiency, behavior disorder with temper tantrums, and sleep and temperature regulation disturbances. However, the few published and anecdotal reports of neuropathology show no visible hypothalamic lesion (Cassidy, 1984).

The first objective finding concerning the etiology of PWS was reported by Ledbetter et al (1981), who used the then newly developed technique of high resolution chromosome analysis to uncover a small interstitial deletion in chromosome 15 long arm in some patients with PWS, but not all. Subsequently, many workers studied series of patients to further delineate the exact location of the de novo deletion (15q11-q13) and the proportion of patients with clinically typical PWS who demonstrated this deletion (series showed from 50% to 100% had a deletion). Recent summaries of these series suggest that approximately 60% - 70% of clinically typical patients have the deletion (Ledbetter and Cassidy, 1988; Butler, 1990). Several of the published series also sought clinical differences between those patients with and without the deletion. Those using careful criteria for diagnosis showed that only the proportion with hypopigmentation differed, with more deleted patients having fair coloring than undeleted, though there were hypopigmented patients in both groups (Butler, 1990).

The report of an occasional family in which more than one child was affected with PWS (Lubinsky et al, 1987) suggests that there may be a recurrence risk which is more than negligible. However, there has never been a reported recurrence in a family in which the proband has a cytogenetic deletion. Suggested recurrence risks are very low (negligible if a deletion is present, less than 1/1000 if it is not) (Cassidy, 1987).

Although parental chromosomes are normal when a deletion is found in a child with PWS, Butler and Palmar (1983) used chromosomal polymorphisms to identify the fact that the deletion occurs in the **paternally derived** chromosome 15. This consistent finding has subsequently been confirmed in several reports (reviewed in Butler, 1990), and more recently has been confirmed using molecular genetic technology (Nicholls, 1989). Although the reason for this was obscure, it became even more intriguing when it was recognized that a proportion of patients with a clinically very distinct disorder, Angelman syndrome, had the same or very similar deletion, and that with Angelman syndrome the deletion occurs in the maternally derived chromosome 15 (Magenis et al, 1987; Donlon, 1988; Knoll et al, 1989).

Angelman syndrome is a relatively newly recognized condition (Angelman, 1965). It is marked by severe mental retardation with absence of speech, ataxic gait with jerky movements, hyperactivity, seizure disorder with characteristic

EEG changes (high amplitude spike and slow waves), and frequent paroxysms of laughter. There are also characteristic dysmorphic features including microbrachycephaly, maxillary hypoplasia, large mouth with widely spaced teeth and frequent tongue protrusion, and prognathism. Interestingly, hypopigmentation, particularly of the choroid and iris, are common. Further delineation of this disorder is currently in progress (some was presented at this workshop), but it is apparent that this disorder is clinically easily distinguished from PWS. Approximately half the patients with Angelman syndrome have a deletion of 15q11-q13 on high resolution cytogenetic analysis (Knoll et al, 1989).

The reason why a deletion in the paternally derived member of a chromosome causes one disorder and in the maternally derived chromosome causes a different disorder is unknown. However, similar phenomena in animals has been called **genomic imprinting** or **genetic imprinting**. This is a process of epigenetic modification which differs in different sexes. Studies are ongoing to help delineate mechanisms of imprinting and explain such different outcomes of differing modification of the same genetic information.

There remains the puzzle of why some patients with PWS (and some with Angelman syndrome) have a detectable cytogenetic deletion and others do not. With the use of DNA markers for the 15q11-q13 region, it was determined that some of these nondeleted patients have a submicroscopic deletion. More interestingly, most without a visible deletion have been found to have two maternally derived chromosome 15q regions and no paternally derived 15q, a phenomenon called **disomy** (Nicholls et al, 1989b). Cases with the two different maternal chromosome 15s (heterodisomy) and with one duplicated chromosome 15 (isodisomy) have been identified, and are detailed in this workshop. Although paternal disomy has been identified in Angelman syndrome, disomy is not as frequent as in PWS (Knoll et al, 1991).

Thus, absence of paternal contribution at proximal 15q

causes the complex features of PWS, and absence of the maternal contribution to the same region causes the many findings in Angelman syndrome. The mechanisms by which these features are produced is unclear. Our knowledge of PWS has increased significantly since the first description of this complex condition in 1956. We have also learned a great deal about its genetics. But there is still much more to be learned. The following workshop presentations describe many of the advances, and the research they represent has contributed greatly to our understanding of this difficult and fascinating disorder.

REFERENCES

Angelman H (1965). "Puppet" children: A report on three cases. Dev Med Child Neurol 7:681-683, 1965.

Aughton DJ and Cassidy SB (1990). Physical features of Prader-Willi syndrome in neonates. Am J Dis Child 144: 1251-1254.

Bray GA, Dahms WT, Swerdloff RS, Fiser RH, Atkinson RL, Carrel RE (1983). The Prader-Willi syndrome. Medicine 62:59-80.

Butler MG (1990). Prader-Willi syndrome: Current understanding of cause and diagnosis. Am J Med Genet 35:319-332.

Butler MG, Meaney FJ, and Palmer CG (1987). Clinical and cytogenetic survey of 39 individuals with Prader-Willi syndrome. Am J Med Genet 26:445-455.

Butler MG and Palmar CG (1983). Parental origin of chromosome 15 deletion in Prader-Willi syndrome. Lancet 1:1285-1286.

Cassidy SB (1984). Prader-Willi syndrome. Curr Probl in Ped 14:1-55.

Cassidy SB (1987). Recurrence risk in Prader-Willi syndrome (Letter). Am J Med Genet 28:59-60, 1987.

Donlon TA (1988). Similar molecular deletions on chromosome 15q11.2 are encountered in both the Prader-Willi and Angelman syndromes. Hum Genet 80:322-328.

Greenswag LR (1987). Adults with Prader-Willi syndrome: A survey of 232 cases. Dev Med Child Neurol 29:145-152.

Holm VA (1981). The diagnosis of Prader-Willi syndrome. In Holm VA, Sulzbacher S, and Pipes PL (eds), Prader-Willi Syndrome. Baltimore, University Park Press, ch. 3.

Holm VA and Pipes PL (1976). Food and children with Prader-Willi syndrome. Am J Dis Child 130:1063-1067.

Knoll J, Nicholls R, Magenis R, Graham J Jr (1989). Angelman and Prader-Willi syndrome share a common chromosome 15 deletion but differ in parental origin of the deletion. Am J Med Genet 32:285-290.

Knoll JH, Glatt KA, Nicholls RD, Malcolm S, and Lalande M (1991). Chromosome 15 uniparental disomy is not frequent in Angelman syndrome. Am J Hum Genet 48:16-21.

Ledbetter, D.H. and Cassidy, S.B. (1900). Etiology of Prader-Willi syndrome. In Caldwell ML and Taylor RL (eds.), Prader-Willi Syndrome: Selected Research and Management Issues. Springer-Verlag, New York, p. 13-28.

Ledbetter DH, Riccardi VM, Airhart SD, Strobel RJ, Keenen SB, and Crawford JD (1981). Deletions of chromosome 15 as a cause of the Prader-Willi syndrome. N. Engl J Med 304:325-329.

Lubinsky M, Zellweger H, Greenswag L, Larson G, Hansmann I, and Ledbetter D (1987). Familial Prader-Willi syndrome with normal chromosomes. Am J Med Genet 28:37-43.

Magenis RE, Brown MG, Lacy DA, Budden S, and LaFranchi S (1987). Is Angelman syndrome an alternate result of del(15)(q11q13)? Am J Med Genet 28:829-838.

Nicholls RD, Knoll JH, Glatt K, Hersh JH, Brewster TD, Graham JM, Wurster-Hill D, Wharton R, Latt SA (1989). Restriction fragment length polymorphisms within proximal 15q and their use in molecular cytogenetics and the Prader-Willi syndrome. Am J Med genet 33:66-77.

Nicholls RD, Knoll JHM< Butler MG, Karam S, Lalande M (1989b). Genetic imprinting suggested by maternal heterodisomy in nondeletion Prader-Willi syndrome. Nature 342:281-285.

Prader A, Labhart A, and Willi H (1956). Ein Syndrom von Adipositas, Kleinwuchs, Kryptorchismus and oligophrenie nach Myotonicartigem Zustand in Neugeborenalter. Schweiz Med Wochenschr 86:1260-1261.

Uehling D (1980). Cryptorchidism in the Prader-Willi syndrome. J Urol. 124:103-104.

Whitman BY and Accardo P (1987). Emotional symptoms in Prader-Willi syndrome adolescents. Am J Med Genet 28:897-905.

Zellweger H (1981). Diagnosis and therapy in the first phase of Prader-Willi syndrome. In Holm VA, Sulzbacher S, and Pipes PL (eds), Prader-Willi Syndrome. Baltimore, University Park Press, ch. 5.

MICRODISSECTION AND MOLECULAR ANALYSIS OF PROXIMAL 15q

Karin Buiting
Institut für Humangenetik
Universitätsklinikum Essen
Hufelandstrasse 55
4300 Essen
Germany

Valerie Greger[1], Ingrid Hörstmann[1], Hermann-Josef Lüdecke[1], Gabriele Senger[2], Uwe Claussen[2], Bernhard H. Brownstein[3], David Schlessinger[3], Joan H.M. Knoll[4], Marc Lalande[4], Bernhard Zabel[5], and Bernhard Horsthemke[1].

Institut für Humangenetik, Universitätsklinikum, Essen, Germany[1]; Institut für Humangenetik, Universität Erlangen, Germany[2]; Department of Genetics, Washington University, St. Louis, Missouri, USA[3]; Children's Hospital, Boston, Massachusetts, USA[4]; and Universitatskinderklinik, Mainz, Germany[5].

INTRODUCTION

The identification of the genetic defect in Prader-Willi syndrome (PWS) and Angelman syndrome (AS) requires the isolation of molecular probes for the critical region (15q11.2-13). In 1986, Donlon et al. constructed a DNA library from a flow sorted inv dup(15)(pter-q13:q13-pter). Clones from this library were instrumental in defining the shortest region of overlap in PWS and AS (Knoll et al, 1990) and in identifying uniparental disomy in non-deletion PWS (Nicholls et al, 1989) and AS (Malcolm et al, 1991).

To obtain more DNA markers for 15q11-13, we used microdissection and enzymatic DNA amplification (Buiting et al, 1990). Several microclones for the critical region have been isolated. Microclone PW71 (D15S63) is deleted in all patients with cytogenetically visible deletions. Microclone MN7 (D15F37S1, D15F37S2, D15F37S3, D15F37S4, D16F37S5, and D15F37S6) detects a variable number of restriction fragments.

NATO ASI Series, Vol. H 61
Prader-Willi Syndrome
Edited by S. B. Cassidy

PREVIOUS AND CURRENT WORK

GTG banded metaphase chromosomes were dissected with the help of a micromanipulator and extended siliconized glass needles. Starting from 20 chromosome fragments, 5000 recombinant clones were obtained. Of these clones, 39% identify single copy human DNA sequences, most of which are derived from the dissected region. Several clones mapping to the critical PWS/AS region were identified. Microclone PW71 detects a single copy sequence in the shortest region of deletion overlap in PWS and AS. MN7 detects a variable low copy number repeat, which we have mapped to 15q11-13 and 16p11.2 by somatic cell genetics and chromosome in-situ hybridization.

Using PCR, we have isolated one yeast artificial chromosome (YAC) for PW71 (125kb) and eight YACs for MN7 (200-600kb) from the YAC library at the Center for Genetics in Medicine (Washington University, St. Louis) (Brownstein et al, 1989; Green et al, 1990). Interspersed repetitive sequence PCR, restriction enzyme analysis, and partial sequencing were used to identify overlaps between the YACs. Two MN7 YACs map to 16p. The other MN7 YACs form three groups of overlapping clones, which contain at least three different MN7 repeat units and cover 1.5 million base pairs of 15q11-13. The repeat units differ by a low frequency of nucleotide exchanges. One YAC which contains two MN7 loci is unstable in yeast cells.

In the mouse, MN7 detects a single locus on chromosome 7, which is transcribed into an mRNA of 11 kilobases (Chaillet et al, 1991; Nicholls et al, 1991). We have isolated two partial cDNA clones from a human placenta cDNA library. Sequence analysis has revealed that one cDNA is from chromosome 15 and the other one from chromosome 16. The function of the MN7 gene family is unknown, but the present of multiple copies of this gene family in proximal 15q may be causally related to the instability of this region and to associated disorders.

We are currently screening the CEPH YAC library (Albertsen

et al, 1989) with other microclones as well as with the flow sorted clones mentioned above. Our goal is to construct a contiguous set of overlapping YAC clones that cover the PWS/AS region.

(Supported by the Deutsche Forschungsgemeinschaft)

REFERENCES

Albertsen HM, Abderrahim H, Cann H, Dausset J, LePaslier D, Cohen D (1990) Construction and characterization of a yeast artifical chromosome library containing seven haploid human genome equivalents. Proc Natl Acad Sci USA 87:4256-4260

Brownstein BH, Silvermann GA, Little RD, Burke DT, Korsmeyer SJ, Schlessinger D, Olson MV (1989) Isolation of single copy human genes from a library of yeast artificial chromosome clones. Science 244:1348-1351

Buiting K, Neumann M, Lüdecke HJ, Senger G, Claussen U, Antich J, Passarge E, Horsthemke B (1990) Microdissection of the Prader-Willi syndrome chromosome region and identification of potential gene sequences. Genomics 6:521-527

Chaillet JR, Knoll J. Hosthemke B, Lalande M (1991) The syntenic relationship of the critical deletion region for the Prader-Willi/Angelman syndromes and proximal mouse chromosome 7. Submitted

Donlon TA, Lalande M, Wyman A, Bruns G, Latt SA (1986) Isolation of molecular probes associated with the chromosome 15 instability in the Prader-Willi syndrome. Proc Natl Acad Sci USA 83:4408-4412 (correction 83:6964)

Green DE and Olson MV (1990) Systematic screening of yeast artificial chromosome libraries by use of the polymerase chain reaction. Proc Natl Acad Sci USA 87:1213-1217

Knoll JHM, Nicholls RD, Magenis RE, Graham JM Jr, Lalande M, Latt SA (1990) Angelman and Prader-Willi syndromes share a common chromosome 15 deletion but differ in parental origin of the deletion. Am J Med Genet 32:285-290

Malcolm S, Clayton-Smith J, Nichols M, Robb S, Webb T, Armour JAL, Jeffreys AJ, Pembrey ME (1991) Uniparental paternal disomy in Angelman's syndrome. Lancet 337:694-697

Nicholls RD, Knoll JHM, Butler MG, Karam S, Lalande M (1989) Genetic imprinting suggested by maternal heterodisomy in nondeletion Prader-Willi syndrome. Nature (London) 342:281-285

Nicholls RD, Horsthemke B, Neumann P (1991) Mouse chromosome mapping of clones from the PWS/AS genetic region. Mouse Genome 89:254

THE IRREGULAR INHERITANCE OF ANGELMAN SYNDROME AND PRADER-WILLI SYNDROME

Marcus E. Pembrey
Mothercare Professor of Pediatrics
Division of Pediatric Genetics
Institute of Child Health
30 Guildford Street
London, England

J. Clayton-Smith, T. Webb, and S. Malcolm
Institute of Child Health, London, England

INTRODUCTION

The clinically distinct developmental disorders, Angelman syndrome (AS) and Prader-Willi syndrome (PWS), can both arise from a de novo deletion of 15q11-13 (Ledbetter et al, 1981; Kaplan et al, 1987; Magenis et al, 1987; Pembrey et al, 1989; Donlon, 1988). However, in AS it is the maternal 15 that is deleted, while in PWS it is the paternal 15 (Butler and Palmer, 1983; Knoll et al, 1989). These observations, plus the fact that AS can be caused by uniparental paternal disomy (Malcolm et al, 1991) and PWS by uniparental maternal disomy (Nicholls et al, 1989), suggest that these loci are subject to genomic imprinting. Normal development cannot be rescued by the inheritance of two intact 15s from one parent. An important distinction between AS and PWS is that affected siblings with AS, although unusual, are considerably more common than familial PWS, and to date patients with familial AS do not have cytogenetic deletions (Pashayan et al, 1982; Fisher et al, 1987; Baraitser et al, 1987; Willems et al, 1987; Fryns et al, 1989; Imaizumi et al, 1990; Knoll et al, 1989).

In PWS there appear to be three mechanisms associated with development of the syndrome:

a) De novo deletion of paternal 15q11-13 (Ledbetter et al, 1981; Butler et al, 1983; Knoll et al, 1989)
b) Uniparental maternal disomy due to meiotic non-

NATO ASI Series, Vol. H 61
Prader-Willi Syndrome
Edited by S. B. Cassidy

disjunction (Nicholls et al, 1989)

c) Rare familial chromosomal aberrations involving 15 (Hasegawa et al, 1984; Hulten et al, 1991)

The first and second can be expected to be associated with low sibling recurrence risks and the third is both rare and usually obvious on cytogenetic analysis of the index patient. Thus, overall, the sibling risk is expected to be low, certainly less than 1%. This is supported by empirical observations (Cassidy, 1987).

Our interest at the Institute of Child Health (ICH), London, is currently focused on the inheritance of AS. The preliminary results of our studies (as of 4/30/91) are presented with a view to defining the different mechanisms involved and associated sibling recurrence risk figures to be used in genetic counseling.

ASCERTAINMENT OF SUBJECTS

By May, 1991, the ICH study on AS included 84 patients from 78 families. The five multiplex families (four sib pairs and one family with three affected children) were actively sought out and include one family from The Netherlands (through the kind cooperation of Dr. I. Dijkstra) and one from Hong Kong (through the kind cooperation of Dr. S. Lam). The English sib pair (Fisher et al, 1987) have to date declined further investigation and are not included in the ICH series. Thus, four multiplex families are known to exist in England. The simplex families, who are all resident in Britain, were ascertained through the Hospitals for Sick Children, Great Ormond Street, self referrals from the British Angelman Syndrome Support Group (where the help of Mrs. Sheila Woolven has been most valuable), and referrals from British clinical geneticists and pediatricians who know of our research interests. We believe the ascertainment is not particularly biased towards those with a known cytogenetic deletion, but we know it is not yet complete, there being over 100 patients known to the Support Group.

With very few exceptions, all the patients in the ICH series have been examined by Dr. Jill Clayton-Smith and the clinical diagnosis was supported by a characteristic EEG (Boyd et al, 1988).

METHODS

Cytogenetic and DNA analysis are as previously published (Malcolm et al, 1991) and the following probes have been used: IR39 (D15S18), ML34 (D15S9), IR43R (D15S11), 189-1 (D15S13), 3.21 (D15S10), IR10-1 (D15S12) - gifts of Dr. M. Lalande, cMW-1 (D15S24) - gift of Dr. D. Ledbetter, and cMS620 (D15S86) - gift of Professor A. Jeffreys.

RESULTS

Cytogenetic deletions

It has been our experience that some patients referred with a previous "routine" cytogenetic analysis indicating no 15q deletion may reveal a deletion of 15q11-13 on examination of high resolution preparations of at least 850 bands. 56 patients (50 families) have been examined at high resolution by T.W. and 32 patients (57%) have a cytogenetically detectable deletion. None of the patients in the five multiplex families showed a deletion, so 64% of index patients of families have a deletion. This compares with 46/84 (55%) of patients and 46/78 (59%) of families in the total ICH series to date.

Parental origin of de novo deletions of 15q11-13

We have determined the parental origin of the deleted chromosome in those families informative for either short arm cytogenetic polymorphisms or restriction fragment length polymorphisms within the deleted region. As of May, 1991, 25 families were informative and in all 25 the deleted chromosome was maternal in origin (Clayton-Smith et al, 1991b).

Uniparental paternal disomy (UPD)

We have undertaken an initial screen for "whole chromosome 15" uniparental disomy using the multi-allelic probe cMS620 (D15S86) which maps to the end of the long arm of 15. As expected, the only class of patients in which UPD was found were the 27 "sporadic" patients with no detectable deletion. Four were uninformative, UPD at D15S86 was excluded in 21, and UPD was found in two patients. These two have been previously reported (Malcolm et al, 1991). One showed paternal heterodisomy, the other paternal isodisomy, and both had typical AS. UPD has been excluded in all 11 patients in the five multiplex families.

Multiplex Families: Inheritance of the 15pter-15q13 Region by Affected Siblings

To test the hypothesis that familial AS may be due to autosomal recessive inheritance of mutations at 15q11-13, we performed linkage studies on four sibships with two children with AS and one sibship with three children with AS. While the appropriate method of sib pair analysis in such a set is still a subject of debate, it can be argued that the family with three affected should count as three rather than two "sib pairs" (L. Sandkuijl, personal communication), making a total of seven "sib pairs". The full details of these families are the subject of a submitted publication and just the overall conclusions are presented here. Linkage analysis using short arm cytogenetic markers and 15q11-13 polymorphic probes indicates that 7/7 affected sib pairs inherited the same maternal chromosome 15 for at least one DNA marker from the cytogenetic short arm markers to D15S24 in 15q13, while 4/7 sib pairs inherited a different paternal chromosome 15, by the same criterion. This effectively excludes an autosomal recessively inherited mutation at 15q11-13, but raises the possibility of maternal transmission of a mutation at this locus.

Other Chromosomal Aberrations Involving Chromosome 15

We believe three families have a rearrangement of chromosome

15 that appears to involve the 15q11-13 region. One of the four such families referred to in our earlier report (Clayton-Smith et al, 1991a) is now thought not to involve a true chromosomal rearrangement, which illustrates the difficulties that can arise in cytogenetic interpretation. One of the three is a pericentric inversion, inv(15)(p11q13), in mother and son, reported earlier (Pembrey et al, 1989). It is now known (thanks to a DNA analysis by Dr. J. Knoll, Boston) that the son has a large deletion. The deletion has arisen de novo, as the mother is heterozygous for a probe deleted in the boy. In another of the families, the mother of the child with AS has a brother with a child with PWS. The nature of the cytogenetic and molecular events in this family are under investigation. The ICH series does not, to date, include the English family with one AS and two PWS patients reported recently (Hulten et al, 1991).

DISCUSSION

The exclusive maternal origin of the de novo deletions of 15q11-13 and the two AS patients with uniparental paternal disomy confirm that the AS locus is normally subject to genomic imprinting, with the paternal allele being silenced during a significant part of brain development. This being so, it is only possible for the AS and PWS locus transcripts to be one and the same if there were some extraordinary switching from paternal to maternal allele silencing during different phases of brain development or in different physical domains of the developing brain. Much more likely is that the AS and PWS loci are separate but close, and this in turn suggests a simple hypothesis for the familial cases of AS. If the mother of affected children with AS carried a paternally derived deletion or point mutation that did not extend to also include the PWS locus, then she herself would be phenotypically normal but would face a 50% chance of any child of hers having AS. This hypothesis for familial AS also explains the absence of cytogenetically detectable deletions in all reported familial cases to date, as well as the paucity of normal siblings in such families. There are no normal siblings in the five multiplex families in the ICH series, which is more in favor of a 50% recurrence risk than 25%.

The data presented by Niikawa, Hamabe and other colleagues at this Workshop (Hamabe et al, 1991) provides direct support for the above hypothesis of autosomal dominant inheritance, modified by genomic imprinting. Three affected siblings, their unaffected mother and maternal grandfather all carried a small deletion detected solely by probe 3.21 (D15S10).

If one accepts the above explanation for all familial AS, then one faces the question as to why the same mechanism does not result in more familial PWS cases than are seen. In PWS, the de novo mutational event would have to arise during oogenesis leading to a carrier male. Differences in the probability of localized mutational events between oogenesis and spermatogenesis is the most likely explanation.

The discoveries of the last few years concerning the complex inheritance of AS allows a tentative conclusion of sibling recurrence risks depending on the genetic mechanism involved in the index case of the family. As far as the families and those providing genetic counseling are concerned, any guidance is better than none, so we do think it is a useful exercise even though further information may lead to some adjustment of the figures in the future. These estimates of the sibling recurrence risks are summarized in Table I.

In view of the fact that cytogenetically visible deletions have yet to be seen in any familial AS cases where the parents' chromosomes are normal, it seems reasonable to give a low recurrence risk in the situation of a regular de novo 15q11-13 deletion. It would be important to exclude any familial predisposing chromosomal rearrangement by high resolution cytogenetics.

Table I. Angelman Syndrome Sibling Recurrence Risks

De novo deletion on cytogenetics with normal parental chromosomes	<2%
Deletion or uniparental disomy in family with rearranged chromosome 15	? 5-10%
Uniparental disomy with normal parental chromosomes	? 2%
Sporadic non-deletion, uniparental disomy excluded	? 10-20%
Familial non-deletion	? 50%

Where such a chromosomal aberration exists, one would be wise to give an intermediate offspring risk to anyone who carried the rearrangement. The empirical data are scant, but recurrences of AS and families with AS and PWS individuals have been reported in extended families carrying chromosomal rearrangements. The affected children may be caused by deletions or uniparental disomy.

The non-disjunction events that underlie uniparental disomy in parents with normal karyotypes have only a small probability of recurring and so uniparental paternal disomy should be associated with a low recurrence risk. Again, familial chromosome arrangements involving 15 could lead to a male producing more than one child with AS with uniparental disomy because of predisposition to meiotic non-disjunction.

Obviously, where a couple have already had two children with AS, the recurrence risk is high. On the current hypothesis discussed earlier, the recurrence risk would be 50%.

The most difficult estimate concerns the risk following a single child with AS who has no detectable deletion and where uniparental disomy is excluded. What must be emphasized in genetic counseling is that artificial insemination by donor (AID) is most unlikely to reduce the risk, since there is no evidence for autosomal recessive inheritance. If the only mechanism for familial AS (in the absence of a familial chromosomal rearrangement) is a mutation confined to the AS locus carried by the mother who inherited it from her father, then gene tracking within the family with linked probes (once these are characterized) should reveal that the chromosome 15 transmitted to the child with AS by the mother would be the one from her own father. Thus, in the future, it may be possible to classify this last group of sporadic cases into those carrying such a grandpaternal chromosome and those where the maternally derived chromosome 15 in the affected child came from the grandmother. The latter would be presumably at a lower risk of recurrence. However, at present, some overall risk figure is needed.

A very crude estimate can be made as follows: there are four affected sibships known in England, three of which are in the ICH series, including the family with three affected children. For the purpose of the calculation, we could say the ICH series contains four sib pairs and 25 sporadic cases where there is no deletion or UPD. We would have to assume a sibship size of two and that there is no significant ascertainment bias towards multiplex families. If this overall group comprises a) families facing a 50% recurrence risk because mother carries an AS mutation and b) families facing a low (2%) risk because the mutation arose *de novo* in the mother, then some overall recurrence risk can be estimated. With respect to class a), for every sib pair where both are affected, there should be two where just one is affected. This, therefore, accounts for eight of the "sporadic" cases, leaving 17 sporadic cases to be assigned to class b) families. In round terms, one third of the overall group is facing a 50% risk and two thirds a 2% risk, so the pooled risk is about 1/6 or 17%. There are, of course, many assumptions in all this and likely ascertainment bias in favor of multiplex families, but it seems reasonable to quote a 10-20% risk until more information is available.

REFERENCES

Baraitser M, Patton M, Lam STS, Brett EM, Wilson J (1987) The Angelman (happy puppet) syndrome: Is it autosomal recessive? Clin Genet 31:323-330

Boyd SG, Harden A, Patton MA (1988) The EEG in early diagnosis of Angelman's (happy puppet) syndrome. Eur J Pediatr 147:508-513

Butler MG, Palmer CG (1983) Parental origin of chromosome 15 deletion in Prader-Willi syndrome. Lancet 1:1285-1286

Cassidy SB (1987) Letter to the Editor: Recurrence risk in Prader-Willi syndrome. Am J Med Genet 28:59-60

Clayton-Smith J, Webb T, Pembrey ME, Malcolm S (1991a) Genetic mechanisms and recurrence risks in Angelman syndrome. J Med Genet 28:568 (abstract)

Clayton-Smith J, Webb T, Pembrey ME, Nichols M, Malcolm S (1991b)Maternal origin of deletion 15q11-13 in 25/25 cases of Angelman syndrome. Hum Genet (in press)

Donlon TA (1988) Similar molecular deletions on chromosome 15q11.2 are encountered in both the Prader-Willi and Angelman syndromes. Hum Genet 80:322-328

Fisher JA, Burn J, Alexander FW, Gardner-Medwin D (1987) Angelman (happy puppet) syndrome in a girl and her brother. J Med Genet 24:294-298

Fryns JP, Kleczkowska A, Decock P, van den Berghe H (1989) Angelman syndrome and 15q11-13 deletions. J Med Genet 26:538-540

Hamabe J, Kuroki Y, Imaizumi K, Sugimoto T, Fukushima Y, Yamaguchi A, Izumikawa Y, Niikawa N (1991) DNA deletion and its parental origin in Angelman syndrome patients. Am J Med Genet (in press)

Hasegawa T, Hara M, Ando M, Osawa M, Fukuyama Y, Takahasi M, Yamada K (1984) Cytogenetic studies of familial Prader-Willi syndrome. Hum Genet 65:325-330

Hulten M, Armstrong S, Challinor P, Gould C, Hardy G, Leedham P, Lee T, McKeown C (1991) Genomic imprinting in an Angelman and Prader-Willi translocation family. Lancet 338:638-639

Imaizumi K, Takada F, Kuroki Y, Maritome K, Hamabe J, Niikawa N (1990) Cytogenetic and molecular study of the Angelman syndrome. Am J Med Genet 35:314-318

Kaplan LC, Wharton R, Elias E, Mandell F, Donlon T, Latt SA (1987) Clinical heterogeneity associated with deletions in the long arm of chromosome 15: Report of 3 new cases and their possible genetic significance. Am J Med Genet 28:45-53

Knoll JHM, Nicholls RD, Magenis RE, Graham JM Jr, Lalande M, Latt SA (1989) Angelman and Prader-Willi syndrome share a common chromosome 15 deletion but differ in parental origin of the deletion. Am J Med Genet 32:285-290

Ledbetter DH, Riccardi VM, Airhart SD, Strobel RJ, Keenan BS, Crawford JD (1981) Deletions of chromosome 15 as a cause of the Prader-Willi syndrome. N Engl J Med 304:325-328

Magenis RE, Brown MG, Lacy DA, Budden S, LaFranchi S (1987) Is Angelman syndrome an alternate result of del(15)(q11q13)? Am J Med Genet 28:829-838

Malcolm S, Clayton-Smith J, Nichols M, Robb S, Webb T, Armour JAL, Jeffreys AJ, Pembrey ME (1991) Uniparental disomy in Angelman syndrome. Lancet 337:694-697

Nicholls RD, Knoll JHM, Butler MG, Karam S, Lalande M (1989) Genetic imprinting suggested by maternal heterodisomy in non-deletion Prader-Willi syndrome. Nature 342:281-285

Pashayan HM, Singer W, Bove C, Eisenberg E, Seto B (1982) The Angelman syndrome in two brothers. Am J Med Genet 13:295-298

Pembrey ME, Fennel SJ, van den Berghe J et al (1989) The association of Angelman's syndrome with deletions within 15q11-13. J Med Genet 26:73-77

Willems PJ, Dijkstra I, Brouwer OF, Smit GPA (1987) Recurrence risk in the Angelman ("happy puppet") syndrome. Am J Med Genet 27:773-780

CHARACTERIZATION OF cDNA CLONES CORRESPONDING TO GENOMIC LOCI REARRANGED IN PATIENTS WITH PRADER-WILLI SYNDROME

C.A. Gregory
Department of Human Genetics
University of Manitoba
250-770 Bannatyne Avenue
Winnipeg, Manitoba R3E 0W3
Canada

A.J. Kirkilionis[1], and J.L. Hamerton[1,2]

Departments of Human Genetics[1] and Pediatrics[2], University of Manitoba, Winnipeg, Manitoba, Canada

INTRODUCTION

Genetic analysis of patients with Prader-Willi syndrome (PWS) began with karyotyping of patients and their parents in the 1970's and continued and expanded in the 1980's with high resolution chromosome banding in an attempt to resolve the type of chromosome rearrangement. The results of high resolution banding in 280 patients with PWS has been summarized (Butler, 1990), and is a deletion of 15q11.2, q11-12, or q11-13 in 56%, a Robertsonian translocation in 0.7%, a reciprocal translocation in 0.7%, an extra isodicentric chromosome 15pter-q11 in 1.4%, other chromosome abnormalities in 0.7%, and a normal karyotype in 41%. Furthermore, the aberrant chromosome 15 was found to be paternal in origin (Butler, 1990). These cytogenetic findings together with phenotype correlation studies showing no phenotypic differences between those PWS patients with or without a cytogenetic deletion, with the exception of hypopigmentation (Butler, 1989), led to the hypothesis that the patients without cytogenetically visible abnormalities may present as PWS as the result of a submicroscopic rearrangement within the region of 15q11-13.

The isolation of genomic DNA probes within the Prader-Willi chromosome region (PWCR) has allowed the testing of this

NATO ASI Series, Vol. H 61
Prader-Willi Syndrome
Edited by S. B. Cassidy

hypothesis. The results of our molecular analysis of patients with PWS (Gregory et al, 1990) are consistent with those reported by others (Donlon, 1988; Nicholls et al, 1989a; Tantravahi et al, 1989) and were concordant with the cytogenetic results in that deletions, and/or duplications were observed in varying combinations. Also, submicroscopic deletions and/or duplications were observed in karyotypically normal patients (Gregory et al, 1990; Gregory et al, 1991). Uniparental disomy of chromosome 15 was shown to be associated with PWS in three patients (Nicholls et al, 1989). The origin of the aberrant chromosome 15 was confirmed by molecular analysis to be paternal (Gregory et al, 1990). These molecular studies have made it clear that rearrangements of loci within this chromosome region are associated with the PWS phenotype.

It has become evident that the only way to determine the etiology of PWS is to isolate the gene or genes involved. The size of this region in kilobases is very large, spanning as much as 6000 kilobases. A long range restriction map of this region spanning pML34, IR10-1, IR4-3 and pTD3-21 using pulse field gel electrophoresis (PFGE) has localized these four probes for proximal 15q to a 2600Kb partial SalI fragment (Kirkilionis et al, 1991). The contrast between the physical distance and the observed recombination frequency suggests that these four loci are located in a recombinational "hot spot". Additionally, PFGE has revealed the presence of HTF (HpaII tiny fragments) islands close to pTD3-21 and at some distance to pML34 (Kirkilionis et al, 1991).

METHODS

Studies were initiated to search for expressed sequences within the PWCR and in particular those genomic loci shown to be rearranged in patients with PWS. The two loci selected (<u>D15S10</u> and <u>D15S0</u>) were used to screen a cDNA human fetal brain library in the cloning vector lambda gt10. Characterization of the cDNA clones involved the following analyses: mapping to a human chromosome, restriction mapping using genomic DNA blots, tissue expression studies using northern blots, conservation of sequences (genomic and cDNA) using a zooblot [genomic DNA from monkey, rat, dog, cow, rabbit, chicken, and yeast] and a primate zooblot [genomic DNA from chimpanzee, orangutan, siamang, Rhesus macaque, King Colobus monkey, and red ruffed lemur], and finally DNA sequencing.

RESULTS

Screening of a fetal brain cDNA library with two genomic DNA probes, pML34 and pTD3-21, which recognize loci rearranged in some patients with PWS (Gregory et al, 1990) has yielded cDNA clones homologous to each genomic locus.

Chromosome mapping, restriction mapping, and sequence conservation

Restriction mapping has verified that each cDNA clone isolated does correspond to its genomic locus. Furthermore, all cDNA clones mapped to chromosome 15 using a southern blot of DNAs from human-hamster hybrid cell lines containing varying combinations of human chromosomes. Conservation studies have shown that conservation of both pML34 and pTD3-21 and their corresponding cDNAs is evident in the representative primate on the blot (rhesus monkey) but not in rat, mouse, dog, cow, rabbit, chicken, or yeast. The hybridization of a primate zooblot with pML34 and pTD3-21 and their cDNAs revealed conservation in chimpanzee, orangutan, siamang, Rhesus macaque, King Colobus moneky, but not in the red ruffed lemus (prosimians).

Expression studies

Northern blots containing mRNA from various cell lines and tissues were hybridized with the genomic DNA probe pML34 and its corresponding cDNAs. Transcripts were detected in Hela cells, HBL 100 normal human epithelial breast with increased amount of c-myc expression, U87.MG human astrocytoma, T986 glioblastoma, H4 neuroglioma, and MCF-7 human breast cancer; no transcripts were detected in JEG human choriocarcinoma, JAR human choriocarcinoma, Bovine adrenal, or IMR32 human neuroblastoma. The same size messages have also been observed in total RNA from normal human lymphoblasts. Further studies revealed no message detectable in northerns containing poly A RNA from adult brain, total RNA from placenta, or RNA from human fibroblasts. Total RNA was also extracted from human hypothalamus and blotted; no message was detected by hybridization of pML34 to the resultant northern blot. Sequential hybridization with actin verified the presence of mRNA on the northern blot.

Expression studies using the cDNAs coresponding to the genomic probe pTD3-21 revealed message of approximately 2.35Kb, 1.9Kb and 1.65Kb when hybridized to northern blots containing mRNA from cell line U87.MG (human astrocytoma). No message was detected from hybridization of a northern blot containing poly A from adult brain or total RNA from human hypothalamus.

Sequencing

Sequencing of the genomic loci pML34 and pTD3-21 and the corresponding cDNAs is continuing, and a Genebank search using sequence thus far generated has revealed no significant homology to known genes. Sequence data from the genomic probe pML34 and one cDNA has identified an exon and a potential open reading frame. The sequence generated from the 3' untranslated region of a pTD3-21 cDNA clone has revealed direct repeats and inverted repeats. It is interesting to note that the corresponding genomic probe has been found to be duplicated in patients with PWS (Gregory et al, 1990; Gregory et al, 1991) and that the presence of the direct and inverted repeats may have some effect on the recombination observed. Additional sequencing is in progress in order to identify the features common to eukaryotic genes and to determine the nature of the coded protein.

DISCUSSION

cDNA clones have been isolated corresponding to loci D15S9

(pML34), which is within the deleted region of patients with PWS, and to locus D15S10 (pTD3-21), which is distal to the deleted region in some deletion patients but found to be duplicated in some karyotypically normal patients (Gregory et al, 1990; Gregory et al, 1991). The genomic and cDNA clones corresponding to loci D15S9 and D15S10 are not conserved in rat, mouse, dog, cow, rabbit, or chicken but are conserved in all primates except the prosimians. Expression studies have shown that while the gene (corresponding to locus D15S9) is expressed in fetal brain it does not appear to be expressed in detectable amounts in adult brain. The size of the message observed in total RNA of cell line U87.MG suggests that two of the cDNAs corresponding to pML34 are not complete cDNAs. A similar situation is evident with the cDNAs corresponding to pTD3-21. These data suggest that the gene(s) involved may be larger than the original genomic cloned segments pML34 and pTD3-21, respectively. Sequence analysis together with the size of the mRNA message will enable us to determine the relationship between the clones corresponding to each genomic locus and whether a full length cDNA is among them.

REFERENCES

Butler MG (1989) Hypopigmentation: a common feature of Prader-Labhart-Willi syndrome. Am J Hum Genet 45:140-146

Butler MG (1990) Review: Prader-Willi syndrome - current understanding of cause and diagnosis. Am J Med Genet 35:316-322

Donlon T (1988) Similar molecular deletions on chromosome 15q11.2 are encountered in both the Prader-Willi and Angelman syndromes. Hum Genet 80:322-328

Gregory C, Kirkilionis A, Greenberg C, Chudley A, Hamerton J (1990) Detection of molecular rearrangements in Prader-Willi syndrome patients using genomic probes recognizing four loci within the PWCR. Am J Med Genet 35:536-545

Gregory CA, Schwartz J, Kirkilionis AJ, Rudd N, Hamerton JL (1991) Somatic recombination rather than uniparental disomy suggested as another mechanism by which genetic imprinting may play a role in the etiology of PWS. Hum Genet (in press)

Kirkilionis A, Gregory C, Hamerton J (1991) Long range restriction map and linkage analysis of the Prader-Willi chromosome region. Genomics 9:524-535

Nicholls R, Knoll J. Glatt K, Hersh J, Brewster T, Graham J Jr, Wurster-Hill D, Wharton R, Latt S (1989a) Restriction fragment length polymorphism within proximal 15q and their use in molecular cytogenetics and the Prader-Willi syndrome. Am J Med Genet 33:66-77

Nicholls R, Knoll J. Butler M, Karam S, Lalande M (1989b) Genetic imprinting suggested by maternal heterodisomy in non-deletion Prader-Willi syndrome. Nature 342:281-285

Tantravahi U, Nicholls R, Stroh H, Ringer S, Neve R, Kaplan L, Wharton R, Wurster-Hill D, Graham J Jr, Cantu E, Frias B, Kousseff B, Latt S (1989) Quantitative calibration and use of DNA probes for investigating chromosome abnormalities in the Prader-Willi syndrome. Am J Med Genet 33:78-87

POSSIBLE GENOMIC IMPRINTING AT THE ANGELMAN SYNDROME GENE LOCUS

Norio Niikawa and J. Hamabe
Department of Human Genetics
Nagasaki University School of Medicine
Sakamoto-Machi 12-4
Nagasaki 852
Japan

SUMMARY

A family in which 3 children were affected with Angelman syndrome (AS) was studied with molecular genetic techniques using 5 different genomic DNA probes each located at 15q11-q12. Southern hybridization with one probe, pTD3-21 (D15S10), revealed one-copy density of the corresponding DNA in all of these children, their phenotypically normal mother and maternal grandfather, but two copies in the father and the maternal grandmother. The D15S10 deletions were confirmed by segregation of RFLPs detected by 3 different restriction enzymes. Southern hybridization for other 4 loci (D15S9, D15S11, D15S12, D15S18) showed the presence of two copies of the sequences in all of these family members. These results were consistent with the genomic imprinting hypothesis for the occurrence of AS, in which the lack of a maternally derived D15S10 allele leads to AS. However, they may not support a model that AS is the alternative phenotype of PWS at the identical locus. This finding also suggests that the presumed AS gene is located near D15S10.

INTRODUCTION

Angelman syndrome (AS) is a mental retardation-malformation syndrome characterized by microcephaly, a large mandible with open mouth and protruding tongue, hypotonia, ataxic gait, easily evoked laughter, seizures, and severe motor and mental retardation. Although AS is clinically distinct from Prader-Willi syndrome

NATO ASI Series, Vol. H 61
Prader-Willi Syndrome
Edited by S. B. Cassidy

(PWS), both syndromes are often related to a similar or identical microdeletion at the 15q11-q12 region (Magenis et al, 1990). The deletion in AS patients was confirmed by molecular studies (Knoll et al, 1989, 1990; Imaizumi et al, 1990). However, the parental origin of the de novo deletion is opposite in the two syndromes: maternal in AS (Knoll et al, 1989; Magenis et al, 1990; Williams et al, 1990) and paternal in PWS (Nicholls et al, 1989a). The genomic imprinting hypothesis has thus been proposed: parental genes located at 15q11-q12 are modified in an epigenetic process, and in the case of AS, paternal gene(s) are imprinted and the loss of the non-imprinted maternally derived locus results in AS phenotype, and vice versa in PWS (Nicholls et al, 1989b; Williams et al, 1990; Knoll et al, 1990).

We report herein the results of a molecular study in an AS family which may support the genomic imprinting mechanism for the occurrence of AS.

MATERIALS AND METHODS

During our molecular study on Japanese AS patients, we encountered an AS family in which 3 children (2 boys, MS and DS, and a girl YS) were affected with AS. High-resolution GTG-banding revealed an apparently normal karyotype in all of them. DNA deletion and RFLP studies were performed on the members of this family using as probes the following 5 cloned DNAs located at 15q11-q12: pML34 (D15S9), pTD3-21 (D15S10), pIR4-3R (D14S11), pIT10-1 (D15S12), and pIR39 (D15S18).

DNA of the family members was prepared either directly from their peripheral blood leukocytes or from EBV-transformed lymphoblastoid cells. They were cut with various endonucleases for different experiments, such as HindIII for probes pML34, pIR10-1, and pTD3-21, and pIR4-3R, ScaI for pML34 and pIR10-1, EcoRV, TagI or AluI for pTD3-21, RsaI or StyI for pIR4-3R, and SacI or BglII for pIR39. DNA fragments were then electrophoresed on 0.5-1.2% agarose gels and blotted onto nylon filters according to the standard methods. The filters were hybridized with ^{32}P-DNA probes labelled by nick translation or the random primer labelling method, washed in stringent conditions at 65° C. in 0.1 x SSC and 0.1% SDS, and were exposed to x-ray film for 24-48 hours. A 4.1-kb PTH-cDNA clone, p20.36, assigned to 11p15.4, was used as an internal standard for densitometry.

Quantitative gene dose analyses were carried out on Southern blots for the 5 loci. An average ratio of the density of an autoradiographic band for each locus to an internal standard band was calculated to determine the copy number of each locus, as described previously (Hamabe et al, 1991a; 1991b). Since such dosage studies are known to have an inherently high error rate, at least 3 independent hybridization experiments for the Southern blotting were performed, and 3 independent autoradiograms for each hybridization were used to confirm the results.

RESULTS

Hybridization with the probe, pTD3-21 (D15S10), revealed a one-copy density of the corresponding HindIII fragments in all the 3 patients (Table I). Enzymes used were BglII for D15S18, HindIII for D15S9, D15S11, and D15S10, and ScaI for D15S12. To confirm their molecular deletions and trace the parental origin of the deletion, RFLP analyses were performed. A probe/enzyme combination, pTD3-21/AluI, gives an RFLP in the Japanese (Kamei et al, 1988), showing 1.5 kb and 0.86 kb + 0.64 kb fragments as A1 and A2 alleles, respectively.

Table I. Relative Density of DNA Fragments at 5 Loci

Patient	D15S18	D15S9	D15S11	D15S10	D15S12
MS	2.1	2.0	2.2	0.9	2.2
YS	1.8	2.1	1.8	1.3	1.7
DS	2.0	2.2	2.0	1.1	1.9

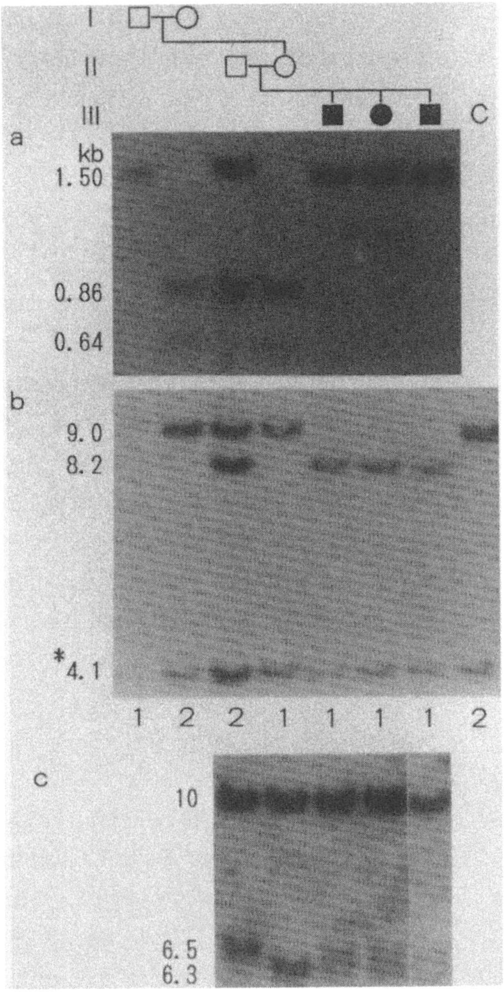

Fig. 1, The pTD3-21/AluI (a), pTD3-21/TagI (b), and pML34/ScaI (c) RFLPs in the family members. C is a phenotypically normal control individual. Asterisk (*) indicates the internal standard PTH DNA. The copy numbers for each family member is shown at the bottom.

Analysis with the combination revealed that each of the three patients was hemizygous for A1, while their father was heterozygous for A1 and A2, the mother hemizygous for A2, the maternal grandfather hemizygous for A1, and the maternal grandmother homozygous for A2 (Fig. 1). Similar transmissions were observed in this family by the analyses with pTD3-21/TaqI and pTD3-21/EcoRV combinations. For the other 4 loci, all of the family members examined had 2 copies of the DNAs (Table II, Fig. 1).

Table II. RFLP Study in Members of the Present AS Family

	Genotype at Loci[a]				
Member	D15S18	D15S9	D15S11	D15S10	D15S12
Maternal grandfather				A1 B1	
Maternal grandmother				A2A2 B1B1	
Father	A2A2 B2B2	A1A1	A1A1 B2B2	A1A2 B1B2 C1C2	A1A1
Mother	A1A2 B2B2	A2A2	A2A2 B2B4	A2 B1 C1	A1A2
Child MS	A2A2 B2B2	A1A2	A1A2 B2B4	A1 B2 C2	A1A1
Child YS	A2A2 B2B2	A1A2	A1A2 B2B4	A1 B2 C2	A1A1
Child DS	A2A2 B2B2	A1A2	A1A2 B2B4	A1 B2 C2	A1A1

[a]D15S18/SacI:A1=14kb/A2=8.5+5.5kb; D15S18/BglIII:B1=9kb/B2=8.5kb; D15S9/ScaI:A1=6.5kb/A2=6.3kb; D15S11/RsaI:A1=1.2kb/A2=1.0kb; D15S11/StyI:B1=3.4kb/B2=2.4kb/B3=1.9kb/B4=0.9kb (Hamabe and Niikawa, 1990); D15S10/AluI:A1=1.5kb/A2=0.86+0.64kb; D15S10/TaqI:B1=9.0kb/B2=8.2kb/B3=8.9kb; D15S10/EcoRV: C1=23.0kb/C2=7.0kb; D15S12/ScaI: A1=17.5kb/A2=16.1kb/A3=12.5kb.

DISCUSSION

A molecular deletion of the D15S10 locus only was detected in all 3 affected children in the present family. Furthermore, the deletion of this locus was also found in their mother and her father (grandfather). Allele transmissions in this family may indicate that in each child an allele that should have been derived from the mother is missing, and the mother herself does not inherit an allele from her father (grandfather). In other words, the D15S10 sequences of the children were all transmitted from the father. Although compound heterozygosity is not absolutely ruled out in the affected children, these data may provide evidence for genomic imprinting, i.e., the deletions derived from females lead to the AS phenotype in offspring and those transmitted from males result in a normal phenotype. Furthermore, since the mother is not affected with AS or PWS, the findings do not support a model that the two syndromes are alternative phenotypes for the identical locus, but rather favor a normal phenotype as the alternative for AS. This in turn suggests different loci for the putative AS and PWS genes. It is conceivable that two different genes exist at the 15q11-q12 region both undergoing genomic imprinting, each of which is responsible for AS or PWS.

The extent of molecular deletions generally differs between AS patients (Knoll et al, 1990; Hamabe et al, 1991a), as seen in PWS patients (Gregory et al, 1990; Hamabe et al, 1991b). Knoll et al (1990) and Hamabe et al (1991a) summarized the deletion ranges in their AS patients and those reported and confined a common deletion site to a segment between D15S11 and D15S10 loci. The finding in the present family suggests the presumed AS gene is located near the D15S10 locus. Thus, a candidate AS gene, if cloned, could be tested in this family.

ACKNOWLEDGMENTS

We express our gratitude to Drs. Y. Kuroki, K. Imaizumi, T. Sugimoto, Y. Fukushima, A. Yamaguchi, Y. Izumikawa, S. Kuwano, R. Matsubayashi and T. Ohtani for referring patients, and to the late Professor S.A. Latt, the American Type Culture Collection (ATCC) for providing probes. This study was supported in part by a grant from the Ministry of Health and Welfare of Japan through Dr. Y.

Nakagome, National Children's Medical Research Center, Tokyo and by a Grant-in-Aid for Scientific Research from the Ministry of Education, Science and Culture.

REFERENCES

Gregory CA, Kirkilionis AJ, Greenberg CR, Chudley AE, Hamerton JL (1990) Detection of molecular rearrangements in Prader-Willi syndrome patients by using genomic probes recognizing four loci within the PWCR. Am J Med Genet 35:536-545

Hamabe J, Niikawa N (1990) A StyI RFLP detected for probe pIR4-3R [D15S11] on chromosome 15q. Nucl Acids Res 18:5579

Hamabe J, Kuroki Y, Imaizumi K, Sugimoto T, Fukushima Y, Yamaguchi A, Izumikawa Y, Niikawa N (1991a) DNA deletion and its parental origin in Angelman syndrome patients. Am J Med Genet (in press)

Hamabe J, Fukushima Y, Harada N, Abe K, Matsuo N, Nagai T, Yoshioka A, Tonoki H, Tsukino R, Niikawa N (1990b) Molecular study of Prader-Willi syndrome: Deletion RFLP and phenotype analyses of 50 patients. Am J Med Genet (in press)

Imaizumi K, Takada N, Kuroki Y, Naritomi K, Hamabe J, Niikawa N (1990) Cytogenetic and molecular study of the Angelman syndrome. Am J Med Genet 35:314-318

Kamei T, Hamabe J, Matsumoto T, Abe K, Harada N, Ishikiriyama S, Hasegawa T, Miyazaki K, Mizuno S, Narahara K, Yukizane S, Niikawa N (1988) A molecular deletion study with Southern hybridization on typical Prader-Willi syndrome (PWS) patients with various chromosome abnormalities involving 15q11-12 and on an atypical PWS patient with apparently normal karyotype. Jpn J Hum Genet 33:477-486

Knoll JHM, Nicholls RD, Magenis RE, Graham JM Jr, Lalande M, Latt SM (1989) Angelman and Prader-Willi syndromes share a common chromosome 15 deletion but differ in parental origin of the deletion. Am J Med Genet 32:285-290

Knoll JHM, Nicholls RD, Magenis RE, Glatt K, Graham JM Jr, Kaplan L, Lalande M (1990) Angelman syndrome: Three molecular classes identified with chromosome 15q11q13-specific DNA markers. Am J Hum Genet 47:149-155

Magenis RE, Toth-Fejel S, Allen LJ, Black M, Brown MG, Budden S, Cohen R, Friedman JM, Kalousek D, Zonana J, Lacy D, LaFranchi S, Lahr M, MacFarlane J, Williams CPS (1990) Comparison of the 15q deletions in Prader-Willi and Angelman syndromes: Specific regions, extent of deletions, parental origin, and clinical consequences. Am J Med Genet 35:333-349

Nicholls RD, Knoll KM, Glatt K, Hersh JH, Brewster TD, Graham JM Jr, Wurster-Hill D, Wharton R, Latt SA (1989a) Restriction fragment length polymorphisms within proximal 15q and their use in molecular cytogenetics and the Prader-Willi syndrome. Am J Med Genet 33:66-77

Nicholls RD, Knoll KM, Butler MG, Karam S, Lalande M (1989b) Genetic imprinting suggested by maternal heterodisomy in non-deletion Prader-Willi syndrome. Nature 342:281-285

Williams CA, Zori RT, Stone JW, Gray BA, Cantu ES, Ostrer H (1990) Maternal original of 15q11-13 deletions in Angelman syndrome suggests a role for genomic imprinting. Am J Med Genet 35:350-353

MOLECULAR ANALYSIS IN ANGELMAN SYNDROME, PRADER-WILLI SYNDROME AND POTENTAL MOUSE MODELS

R.D. Nicholls
Department of Neuroscience, Center for Mammalian Genetics
and Division of Genetics, Department of Pediatrics
University of Florida College of Medicine
Gainesville Florida 32610
USA

and

W. Gottlieb [1,2], M.J. Mascari[4], E.M. Rinchik[4], and G.S. Pai[6], D.J. Driscoll[2,3], M.G. Butler[7], R.T. Zori[2,3], P.E. Neumann[8], M.F. Waters[2,3], J.L. Zackowski[2,3], B. Horsthemke[9], P.K. Rogan[4], R.L. Ladda[4], C.A. Williams[2,3].

Department of Neuroscience[1], Center for Mammalian Genetics[2], and Division of Genetics, Department of Pediatrics[3], University of Florida College of Medicine, Gainesville, FL, U.S.A.; Penn State University College of Medicine, Hershey, PA[4]; Oak Ridge National Laboratory, Oak Ridge, TN[5]; Medical University of South Carolina, Charleston, SC[6]; Vanderbilt University Medical Center, Nashville, TN[7]; Harvard Medical School, Boston, MA[8]; University of Essen, Essen, Germany[9].

INTRODUCTION

Deletions or uniparental disomy for human chromosome 15q11q13 result in Prader-Willi syndrome (PWS) and Angelman syndrome (AS), depending upon the respective parental origin. Clinical and cytogenetic aspects of the two disorders are covered elsewhere in this book. Molecular studies have indicated that the extent of deletions in PWS and AS do not differ (Knoll et al, 1990). That deletions in PWS occurred predominantly on the paternal chromosome 15 was first demonstrated using cytogenetic polymorphisms (Butler et al, 1986), and we now know that all cases are paternal in origin (Nicholls et al, 1989a; Nicholls et al, 1989b; Butler, 1990). In contrast, deletions in AS always occur on the maternal chromosome 15 (Knoll et al, 1989).

More recently, molecular analyses have shown that some PWS patients without a cytogenetic deletion have two copies of a maternal chromosome 15q11q13 (Nicholls et al, 1989b; Mascari et

NATO ASI Series, Vol. H 61
Prader-Willi Syndrome
Edited by S. B. Cassidy

al, 1990). Since both chromosomes derive from only one parent, this is termed uniparental disomy (UPD) (Engel, 1980). A small number of AS patients without a deletion have paternal UPD (Malcolm et al, 1991; Nicholls et al, unpublished). Therefore, PWS originates from the lack of paternal genetic information in 15q11q13 (whether by deletion or UPD) and conversely, AS originates from the lack of maternal alleles in 15q11q13 (whether deletion or UPD). The parental origin effect in PWS and AS provides provocative evidence that genes within 15q11q13, and the nature of the disorder, are controlled by genomic imprinting (Sapienza, 1990).

DNA markers from 15q11q13 have been characterized genetically (Donlon et al, 1986; Nicholls et al, 1989a; Nicholls et al, 1989b; Tantravahi et al, 1989; Mascari et al, 1990) and a genetic and physical map established (see Figure 1; Nicholls et al, in preparation). Using these DNA markers, we have studied a large population of AS patients and 30 cytogenetically non-deleted PWS patients. DNA segments from human 15q11q13 that are conserved in evolution were also used to map the syntenic locations in the mouse genome as an aid to the development of mouse models of AS, PWS, and genomic imprinting.

RESULTS AND DISCUSSION

Deletion and Disomy Studies of Angelman and Prader-Willi Syndrome Patients

DNA markers from chromosome 15q11q13, as well as from near the telomere (Armour et al, 1990), have been utilized in two different studies of AS and PWS. One study population consisted of 27 AS patients and families, identified primarily through the Angelman Research Group registry in the Division of Genetics, University of Florida College of Medicine (Driscoll DJ, Zori RT, Williams CA). In parallel, we studied 30 PWS patients and families identified primarily through Genetics Clinics at Vanderbilt University, Tennessee, USA (Butler MG) and Hershey Medical Center,

Pennsylvania, USA (Ladda RL, Mascari MJ). The second study selected cytogenetically normal PWS patients whereas for the first study the only criteria was clinical diagnosis of AS (and availability of both biological parents). Overall results are shown in Table I and Figure 1.

Table I. Summary of molecular analysis in AS and PWS

(UPD = uniparental disomy; # = 1 pair of monozygous twins, included as a single case.)

AS Patients			**PWS Patients (cytogenetics normal)**		
molecular deletion 13 maternal (7 noninformative)	20	(74%)	molecular deletion 7 paternal (2 noninformative)	9	(30%)
paternal UPD	1	(4%)	maternal UPD	17#	(57%)
biparental inheritance 2 "atypical"	6	(22%)	biparental inheritance all are "atypical"	4	(13%)
TOTAL:	27		TOTAL:	30	

<u>Deletion Studies:</u> Five probes (IR4-3R, 189-1, 34, 3-21, IR10-1) are deleted in nearly all AS and PWS patients with a cytological or molecular deletion (Figure 1, hatched boxes). Two PWS patients, one with a deletion of only the first four markers (Mascari et al, in preparation) and one with an unbalanced translocation, t(Y;15), (Nicholls et al, 1989a; Tantravahi et al, 1989), have probe IR10-1 intact. A dashed line indicates that a marker is deleted in only some patients: deletion of IR39d does not change the PWS or AS phenotype from that seen in typical cases. Similarly, two AS patients deleted for CMW-1 are indistinguishable in phenotype from classical AS, whereas a PWS patient deleted for CMW-1 has a more severe phenotype than classical PWS. The latter results may indicate a more extensive deletion, or alternatively a gene that is expressed only from the paternal chromosome. Note that the probe order is based on these deletion analyses in conjunction with crossover and pulsed field gel electrophoresis (PFGE) data (Nicholls et al, unpublished data).

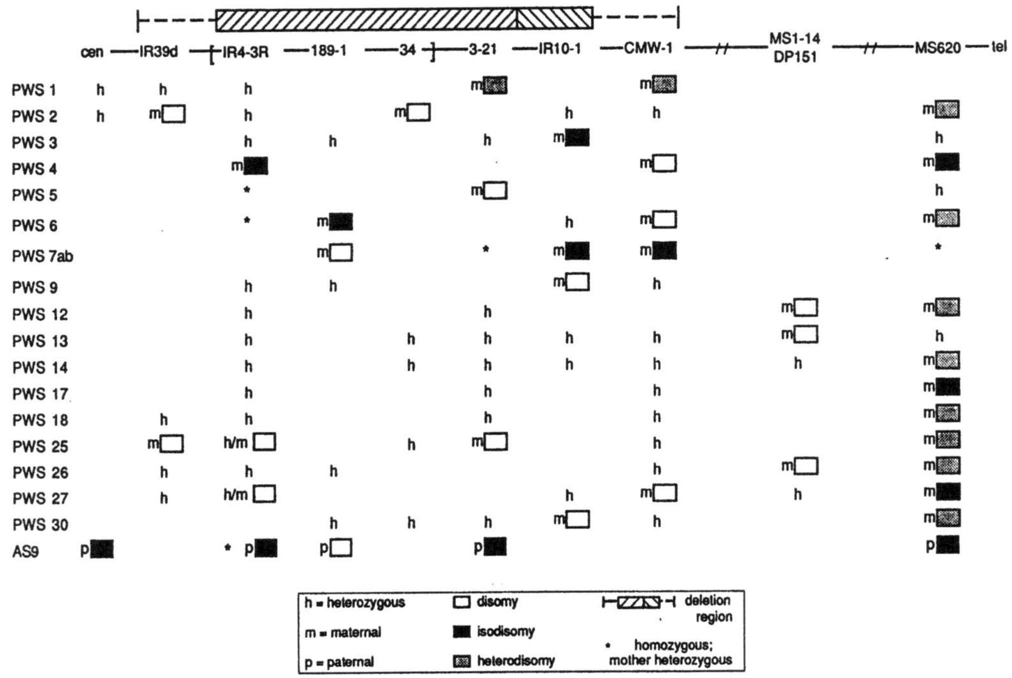

Figure 1: Genetic map of human chromosome 15q11q13, showing deletion regions (top), DNA probe order (line 2), and uniparental disomy (UPD; lines 3-20). Details of deletions and UPD are explained in the text.

The only difference between deletions in AS and PWS patients is the parental origin (Table I), as previously described (Butler et al, 1986; Knoll et al, 1989; Nicholls et al, 1989a; Butler, 1990). Patients within PWS and AS groups can, however, differ from each other with respect to hypopigmentation, which seems to be associated with the presence of a deletion (Butler, 1989; Fryburg et al, 1991). Therefore, the presence of this finding in both patient groups and association with a deletion allows the postulate that a gene involved in pigment formation maps to 15q11q13 and shows a gene dosage effect, but not an imprinting phenomenon. One putative gene in this region has been isolated (DN10) which spans at least 110 kb across the IR10-1 locus: studies in the mouse suggest DN10 may be a candidate for the hypopigmentation gene (see below).

Disomy Studies: One AS patient (identified by Pai GS) was the only case of paternal UPD found (4% of study population, Table I). This low frequency of UPD in AS is similar to that found by Malcolm and co-workers (Malcolm et al, 1991; Pembrey et al, this Workshop). In the PWS study we found 18 patients with maternal UPD (Table I; 17 independent cases with one pair of monozygous twins concordant for PWS). The finding of 9 large deletions in our cytogenetically nondeleted PWS patients (Table I) represent insufficient cytogenetic resolution and not submicroscopic deletions. Nevertheless, combined with the literature estimates of 60-70% for cytogenetic deletions in PWS (Butler, 1990), our findings in 30 "nondeleted" patients allow us to estimate frequencies of 72% deletion, 23% UPD, and 5% biparental inheritance, in PWS. Nearly identical figures have been obtained by Robinson et al (this Workshop): 71%, 22%, and 7%, respectively, for classical patients. We, and Robinson et al, also demonstrated that the few PWS cases with biparental inheritance in our studies were atypical in phenotype, and overall concluded that 1) UPD is very frequent in PWS and 2) molecular diagnosis may be applicable to most patients with PWS.

UPD in PWS and AS has been demonstrated with DNA markers from 15q11q13 and near the telomere (Figure 1), as well as throughout chromosome 15 (Rogan P and Nicholls RD, in preparation). UPD (Figure 1, lines 3-20) can be determined when maternal alleles are different from paternal alleles at a locus. Three distinct inheritance patterns can be detected for UPD. 1) When one parent is heterozygous for a marker, the inheritance of both alleles from this parent, and none from the other parent, is distinctive for heterodisomy (Figure 1, stippled boxes). 2) Inheritance of two copies of only one allele from one parent defines isodisomy (Figure 1, filled boxes). Note that markers must have at least three alleles to identify these situations. 3) If inheritance at a locus is from only one parent, but both parent and child are homozygous, then disomy is established (Figure 1, open box), but iso- or heterodisomy are indistinguishable. Patients may also be heterozygous for a marker but noninformative for parental origin of both alleles. Once UPD has been established for at least one marker on chromosome 15, it is possible to infer (but not prove, as biparental inheritance is a formal possibility) that h represents heterodisomy. Likewise, an asterisk (*) in Figure 1 indicates the possibility of isodisomy, since the patient is homozygous when the relevant parent is heterozygous.

Maternal recombination can be seen in several cases (Figure 1); for example, in PWS3 between 3-21 and IR10-1 and in PWS5 between IR4-3R and the telomere. Similar analyses show recombination in PWS 6, 17 and 27. Additional crossovers are detected when further markers from throughout chromosome 15 are analyzed (Rogan R and Nicholls RD, in preparation). If markers very close to the centromere show heterodisomy then the nondisjunction occurs in meiosis I, while isodisomy is indicative of meiosis II. It is important to find highly polymorphic markers close to the centromere, since meiotic crossover appears to be quite common in 15q11q13 from our molecular studies and the analyses of cytological chiasmata (Saadallah and Hulten, 1983). Nonetheless, it appears that most PWS disomies are due to meiosis I events (Figure 1). Further, maternal ages for the disomy group (mean 30.5 years, sd=7.7, N=17) were significantly elevated (p=0.018 using ANOVA) over maternal ages in a deletion group (mean 24.2 years, sd=4.1, N=11). Paternal ages reflected the mothers' ages. Robinson et al (this Workshop) have obtained similar data on recombination plus nondisjunction in meiosis I, and showed elevated maternal ages for 7 cases of UPD in PWS. These findings are similar to those for tisomy 21 in Down syndrome (Hassold and Jacobs, 1984). The low frequency of paternal UPD in AS may indicate that paternal nondisjunction for chromosome 15 is not as common as maternal nondisjunction, with PWS UPD arising via a trisomic intermediate. Alternatively, it may be that paternal UPD is less viable in utero.

Molecular Map of Chromosome 15q11q13

Three independent but complementary techniques have been used to map DNA markers in 15q11q13, resulting in the probe order shown in Figure 1.

Crossover analysis: First, RFLPs for all markers in 15q11q13 were analyzed for segregation in 3 three generation pedigrees (Nicholls et al, 1987). Three generation analysis allows determination of phase for both alleles in a parent and the subsequent distribution of alleles between homologous chromosomes in the offspring (White and Lalouel, 1988). In this way, recombination can be identified in the third generation. Analysis of crossovers in 22 meioses for both paternal and maternal chromosomes identified 20 crossovers in chromosome 15, with 12 localized between probes in 15q11q13.

Deletion analysis: Analysis of PWS and AS patients with deletions or unbalanced translocations was described above (see Figure 1) and the probe order was consistent with that determined from crossover data.

PFGE: A preliminary long range restriction map is consistent with the order proposed in Figure 1 (Nicholls et al, 1989b; Nicholls et al, in preparation). Most probe and enzyme combinations detect multiple fragments resulting from partial digestion at methylated sites (Nicholls et al, 1989b). Similar results are obtained by Woodage et al (this Workshop) and Kirkilionis et al (Kirkilionis et al, 1991), who report some similar sized fragments, but many differences, precluding map construction from the multiple sets of data. Thus, it is essential to use multiple techniques to map 15q11q13. The map we have obtained (Figure 1) is consistent through all sets of data.

Mouse Mapping Identified Potential Mouse Models for PWS and AS

Evolutionarily conserved clones from human chromosome 15q11q13 were mapped in the mouse genome so as to identify the homologous mouse locus and potential mouse models for the human disorders (Nicholls, 1989; Nicholls et al 1990; Nicholls et al, 1991). In particular, it would be of interest to know whether any known mouse genes, traits (such as obesity, seizures, pigmentation, ataxic gait), or developmental mutants map to the homologous locus.

Two clones from human 15q11q13 were mapped to mouse chromosome 7 using analysis of: (a) RFLP segregation in recombinant inbred (RI) strains; (b) interspecies backcrosses between laboratory mice and M. spretus; and (c) deletion mapping of radiation-induced mutations at the pink-eyed dilution (p) locus, which is a hypopigmentation locus. These studies demonstrated the order: cen---Tam-1---D7Nic2 (MN7), Mtv-1---D7Nic1 (DN10), Mpmv-1, Xmv-33---c---tel. A third clone, cDNA DN34 detects multiple loci in the mouse genome; a polymorphic locus with RI strains maps to mouse chromosome 5 (D5Nic1), but preliminary data suggests a chromosome 7 (D7Nic3) locus.

A chromosome 15q11q13 cDNA, DN10, as well as MN7, is not separable from p in backcrosses and is deleted in most prenatally lethal p-locus deletions and from a viable p-mutation characterized by small size, nervous behaviour, and male sterility. These results indicate that the hypopigmentation observed in PWS and AS could be related to p-locus function in the mouse, and suggest that the developmental phenotypes observed in the p-series of deletions may define a series of mouse models for studying the genetics of PWS and AS (Nicholls RD, Rinchik EM et al, in preparation).

CONCLUSIONS

Seventy to 80% of patients with either AS or PWS have a large molecular deletion of chromosome 15q11q13, very similar in proximal and distal extent and differing only in parental origin. UPD is rare in AS (2-4%) but common in PWS (20-25%). This finding, the advanced maternal age associated with maternal disomy in PWS, and the patterns of recombination in chromosome 15 in these cases, all point toward maternal nondisjunction in meiosis I as the primary mechanism in etiology of maternal disomy, most likely through a trisomic intermediate. Our studies and those of Robinson et al (this Workshop) suggest that molecular diagnosis may be applicable to most patients with PWS. Patients with either syndrome and a large deletion tend to be hypopigmented, whereas non-deletion patients do not, suggesting that a putative gene involved in the pigment system is affected by gene dosage and not by genomic imprinting. A mouse model for this gene and phenotype is the pink-eyed dilution (p) mutation on mouse chromosome 7. Deletions of genes at this mouse locus (homologous to human 15q11q13) in prenatally lethal and other radiation-induced mutations represent potential mouse models for studies of gene expression, phenotypic effect, and genomic imprinting in AS and PWS.

ACKNOWLEDGEMENTS

We are indebted to Karen Avidano, Angela Garcia, Allison Koweek, Shawn Baxley; Dr. Wendy Robinson for discussions; and to all colleagues who provided support, discussion and patients. Dr. Jeffrey Longmate is thanked for help with statistical analyses of parental age. R.D. Nicholls was aided by Basil O'Connor Starter Scholar Research Award No. 5-802 from the March of Dimes Birth Defects Foundation, and is a PEW Scholar in the Biomedical Sciences.

REFERENCES

Armour JAL, Povey S, Jeremiah S. Jeffreys AJ (1990) Systematic cloning of human minisatellites from ordered array charomid libraries. Genomics 8:501-512

Butler MG, Meaney FJ, Palmer CG (1986) Clinical and cytogenetic survey of 39 individuals with Prader-Labhart-Willi syndrome. Am J Med Genet 23:793-809

Butler MG (1989) Hypopigmentation: A common feature of Prader-Labhart-Willi syndrome. Am J Hum Genet 45:140-146

Butler MG (1990) Prader-Willi syndrome: Current understanding of cause and diagnosis. Am J Med Genet 35:319-332

Donlon TA, Lalande M, Wyman A, Bruns G, Latt SA (1986) Isolation of molecular probes associated with the chromosome 15 instability in the Prader-Willi syndrome. Proc Natl Acad Sci USA 83:4408-4412

Engel E (1980) A new genetic concept: Uniparental disomy and its potential effect, isodisomy. Am J Med Genet 6:137-143

Fryburg JS, Breg WR, Lindgren V (1991) Diagnosis of Angelman syndrome in infants. Am J Med Genet 38:58-64

Hassold TJ, Jacobs PA (1984) Trisomy in man. Ann Rev Genet 18:69-97

Kirkilionis AJ, Gregory CA, Hamerton JL (1991) Long-range restriction mapping and linkage analysis of the Prader-Willi chromosome region (PWCR). Genomics 9:524-535

Knoll JHM, Nicholls RD, Magenis RE, Graham JM Jr, Kaplan L, Lalande M (1989) Angelman and Prader-Willi syndromes share a common chromosome 15 deletion but differ in parental origin of the deletion. Am J Med Genet 32:285-290

Knoll JHM, Nicholls RD, Magenis RE, Glatt K, Graham JM Jr, Kaplan L, Lalande M (1990) Angelman syndrome: Three molecular classes identified with chromosome 15q11q13-specific DNA markers. Am J Hum Genet 47:149-155

Malcolm S, Clayton-Smith J, Nichols M, Robb S, Webb T, Armour JAL, Jeffreys AJ, Pembrey ME (1991) Uniparental paternal disomy in Angelman's syndrome. Lancet 337:694-697

Mascari MJ, Ladda RL, Butler MG, Gottlieb W, Nicholls RD (1990) Molecular analysis in non-deletion Prader-Willi syndrome. Am J Hum Genet 47:A228

Nicholls RD, Tantravahi U, Fuller R, Stroh H, Wharton R, Latt SA (1987) Molecular studies of the proximal long arm of human chromosome 15 and the Prader-Willi syndrome. Am J Hum Genet 41:104A

Nicholls RD, Knoll JH, Glatt K, Hersh JH, Brewster TD, Graham JM Jr, Wurster-Hill D, Wharton R, Latt SA (1989a) Restriction fragment length polymorphisms within proximal 15q and their use in molecular cytogenetics and the Prader-Willi syndrome. Am J Med Genet 33:66-77

Nicholls RD, Knoll JHM, Butler MG, Karam S, Lalande M (1989b) Genetic imprinting suggested by maternal heterodisomy in non-deletion Prader-Willi syndrome. Nature 342:281-285

Nicholls RD (1989) A cDNA from poximal human chromosome 15 maps near Mtv-1 on mouse chromosome 7. Mouse Newsletter 84:87-88

Nicholls RD, Horsthemke B, Neumann PE (1991) Mouse chromosome mapping of clones from the PWS/AS genetic region. Mouse Genome 89:254

Saadallah N, Hulten M (1983) Chiasma distribution, genetic lengths, and recombination fractions: A comparison between chromosome 15 and 16. J Med Genet 20:290-299

Sapienza C (1990) Parental imprinting of genes. Sci Am 263:52-06

Tantravahi U. Nicholls RD, Stroh H, Ringer S, Neve RL, Kaplan L, Wharton R, Wurster-Hill D, Graham JM Jr, Cantu ES, Frias JL, Kousseff B, Latt SA (1989) Quantitative calibration and use of DNA probes for investigating chromosome abnormalities in Prader-Willi syndrome. Am J Med Genet 33:78-87

White R, Lalouel J-M (1988) Chromosome mapping with DNA markers. Sci Am 258:40-48

CLINICAL, MOLECULAR, AND CYTOGENETIC SURVEY OF POTENTIAL PRADER-WILLI SYNDROME PATIENTS

W. P. Robinson
Institute of Medical Genetics
Department of Pedaitrics
University of Zürich
Ramistrasse 74
CH8001 Zürich
Switzerland

A. Bottani, X. Yagang, J. Balakrishnan, F. Binkert, M. Mächler, A. Prader, A. Schinzel.

Institute of Medical Genetics and the Department of Pediatrics, University of Zürich, Switzerland

INTRODUCTION

The association of an interstitial deletion of chromosome 15q11-13 with the Prader-Willi syndrome (PWS) has been known since 1981 (Ledbetter et al, 1981 and 1982). However, about 37% of patients are cytogenetically normal (Ledbetter et al, 1987). The discovery of maternal heterodisomy for 15q11-13 in two patients with PWS (Nicholls et al, 1989) raised the question of whether such a phenomenon could explain many or even all non-deletion patients. To answer such a question requires a very good clinical diagnosis and reliable molecular data, and was the major impetus for the work presented here.

In an ongoing study at the University of Zürich, patients referred for genetic analysis as possibly having PWS, by virtue of having some features typical of PWS, have been examined cytogenetically, molecularly and clinically. As the clinical, molecular, and cytogenetic investigations have been done completely independently, this study is also designed to test the utility of molecular probes for routine or early PWS diagnosis, to identify patients who might not completely fulfill diagnostic criteria for PWS but show chromosome 15 abnormalities, and to distinguish clinical features between disomic and deleted patients. Thirty-six patients have been examined, 28 of whom are typical of PWS based

NATO ASI Series, Vol. H 61
Prader-Willi Syndrome
Edited by S. B. Cassidy

upon clinical reevaluation by one or more of us (AP, AS, and AB).

METHODS

Clinical diagnosis of all patients was based on experience and the ten criteria listed in Table I. Only one of these features was allowed to be lacking in a patient considered to be typical of the PWS. If more than three features were absent, the patient was not considered to have the PWS. With 2-3 features lacking he/she was considered to have "atypical PWS". Features were only considered if they could be expected to be present; hence, obesity was disregarded if the patient was younger than 2 years old, hypogenitalism was not scored in females, diminished puberty was not considered in children, etc.. In order to evaluate obesity, we considered both weight percentile and skinfold thickness. Patients with increased skinfold but grossly diminished muscle mass were not always above the 95th percentile in weight but still considered to be obese.

Table I. PWS Diagnostic Criteria

Pregnancy:	Diminished fetal movements
Neonatal period:	Severe hypotonia
	Feeding difficulties
	In boys: hypogenitalism
Early childhood:	Retarded motor/mental development
	Hypotonia
	Increased appetite
	Slender hands and fingers, short feet
	Characteristic facies
Lake childhood:	Obesity with knock-knees
	Diminished puberty
	Predicted adult height below target height

G-banding of prometaphase chromosomes was performed using

standard cytogenetic methods.

Molecular studies on all patients and available parents were undertaken using the polymorphic probes in the PWS region: p39 (D15S18); pML34 (D15S9); pTD3-21 (D15S10); IR4-3 (D15S11); IR10-1 (D15S12); and p189-1 (D15S13). A constant probe from chromosome 7 (pJ3.11) was used as an internal control for dosage comparison using densitometric analysis of autoradiographs. Deletions were identified by comparing dosage of chromosome 15 alleles in patients to heterozygous and homozygous controls. In addition, the probes CMW-1 (D15S24), MS1-14 (D15S1) and MS-620 were used to investigate uniparental disomy. Uniparental disomy was considered when a patient showed only maternal inheritance but did not appear deleted at a particular chromosome 15 locus. Non-paternity was excluded in all families using the Jeffreys hypervariable probe 33.11.

RESULTS

Cytogenetic:

Of the 28 patients identified as fulfilling criteria for PWS, cytogenetically visible interstitial deletions of 15q11.2 were identified in 13 patients, 2 patients had deletions due to unbalanced translocations involving chromosome 15, one patient had a paternally derived ring 15, and one carried a *de novo* apparently balanced t(13;X). The 11 remaining PWS patients and all 8 patients not fulfilling criteria for PWS had no cytogenetically visible aberration.

Molecular genetic:

A deletion of the 15q11.2-q12 region was identified molecularly in 20 of the 28 patients with PWS, including all 15 patients identified cytogenetically as deleted. The origin of the deletion could be shown to be from the father in all 11 cases which were informative. In addition, 6 cases of maternal disomy (4 heterodisomic and 2 isodisomic for 15q11.2-12) were also observed. All of these patients were clinically classified as "typical" of the PWS. Two clinically "atypical" PWS patients showed normal diploid maternal and paternal inheritance for the 15q11.2-12 region, however one carried a ring-15 and the other carried a t(X;13). These findings are summarized in Table II.

Table II. Molecular Results Using Chromosome 15 Probes

	CLINICAL CLASSIFICATION		
	Typical PWS	Atypical PWS	Non-PWS
Deletion	20	0	0
Maternal disomy	6	0	1*
Normal	0	2**	6
Inconclusive	0	0	1

* This patient has been reclassified as typical of the PWS subsequent to clinical reevaluation
**Both these patients have other chromosomal abnormalities

One of the patients originally thought not to fulfill criteria for diagnosis as PWS showed maternal heterodisomy. This patient was born two months premature and although he showed a reduction in intrauterine activity and was severe hypotonia at birth, he did not show the typical facial features of the PWS (but rather showed facial features associated with prematurity) and did not show apparent developmental delay upon clinical evaluation at 8 months of age. He was considered by a pediatric neurologist to have spinal muscular atrophy and a muscle biopsy had been scheduled. Clinical reevaluation of this patient at one year of age and subsequent to the genetic findings, indicates that this patient now shows features typical of the PWS, including facies and mental/motor retardation.

None of the remaining 7 patients not fulfilling criteria for PWS carried a molecular deletion or could be shown to be disomic (in one of these patients disomy could not be excluded because parents were not available).

Five of the 7 patients who had disomy showed clear non-paternal inheritance using probe MS-620 which recognizes a telomeric locus with variable number of tandem repeats (VNTR). This, along with the observation that all genotypes were consistent with maternal disomy for all 9 chromosome 15 probes examined, indicates that they are not disomic only for the 15q11.2-12 region, but for the entire chromosome 15. When considering the entire chromosome, 4 patients showed partial iso- and partial heterodisomy, indicating that normal recombination had taken place in the mother prior to the non-disjunction event, and 3 patients appeared to be heterodisomic for the entire chromosome.

Clinical Comparisons:

The parental ages at the time of birth of the disomic patients were significantly higher than parental ages of deletion patients (Table III). This presumably reflects the increase in maternal age associated with meiotic non-disjunction. No other clear differences between disomic and deletion patients could be found, although sample sizes are still small.

Table III. Mean Parental Ages at Birth

	Mothers	Fathers
Deletion (N=19, 17)	27.4	30.3
Disomy (N=7, 7)	35.6	38.7

P <0.01 for each (Student's t-test)

DISCUSSION

In summary, a deletion of 15q11.2-12 from the father or maternal disomy of the entire chromosome 15 appear to explain the majority of PWS cases and all of those with typical features. Molecular examination should thus provide a reliable diagnostic tool. Molecular probes detected deletions in 71% of patients with PWS, including 5 in whom no cytogenetic aberration was found, thus confirming that molecular studies are more sensitive than cytogenetic ones for detecting deletions.

The striking increase in parental ages is consistent with the hypothesis that disomic patients normally result from either two independent non-disjunction events, one in each parent, or an initial trisomy. As the patients identified as disomic do not show any additional or more severe features than typical deletion patients do, then one must conclude that there is only one imprinted region on chromosome 15 (within 15q11.2-12).

REFERENCES

Ledbetter DH, Greenberg F, Holm VA, Cassidy SB (1987) Conference report: Second annual Prader-Willi syndrome scientific conference. Am J Med Genet 28:779-790

Ledbetter DH, Mascarello JT, Riccardi VM, Harper VD, Airhart SD, Strobel RJ (1982) Chromosome 15 abnormalities and the Prader-Willi syndrome: A follow-up report of 40 cases. Am J Hum Genet 34:278-285

Ledbetter DH, Riccardi VM, Airhart SD, Strobel RJ, Keenan SB, Crawford JD (1981) Deletions of chromosome 15 as a cause of the Prader-Willi syndrome. N Engl J Med 304:325-329

Nicholls RD, Knoll JHM, Butler MG, Karam S, LaLande M (1989) Genetic imprinting suggested by maternal heterodisomy in non-deletion Prader-Willi syndrome. Nature 342:281-285

CYTOGENETIC COMPARISON BETWEEN PRADER-WILLI AND ANGELMAN SYNDROMES

R. Ellen Magenis and SuEllen Toth-Fejel
Department of Molecular and Medical Genetics
Child Development Rehabilitation Center
Oregon Health Sciences University
707 S.W. Gaines Road
Portland, Oregon 97201

INTRODUCTION

The Prader-Willi syndrome (PWS) was recognized as a clinical entity (Prader et al, 1956) over 30 years prior to the finding of the deletion that is now observed in the majority of cases (Ledbetter et al, 1981). The clinical features, which include hypotonia and feeding problems with failure to thrive in infancy, followed by rapid weight gain, evidence of global developmental delay, behavior difficulties in early childhood and the characteristic facial features, appeared to be the same whether a deletion of chromosome 15 was present or not. The facial features of PWS and typical obesity level are illustrated in Fig. 1.

In 1965, Angelman described three unrelated "puppet children" who had flat heads, jerky movements, protruding tongues and bouts of laughter. A patient with "typical" Angelman syndrome (AS) is demonstrated in Figure 2. Though no chromosome abnormalities were detected earlier, in 1987 two different groups, Kaplan et al. and Magenis et al., published their findings in four patients with AS. Several reports followed in 1988 - 1990 which confirmed the original observations (Donlon, 1988; Knoll et al, 1989; Pembrey et al, 1989; Cooke et al, 1989; Williams et al, 1989 a,b; Imaizumi et al, 1990).

Because of the clinical distinctness of AS and PWS, we (Magenis et al, 1990) postulated that the deletions were not identical but adjacent and therefore involved a different set of

NATO ASI Series, Vol. H 61
Prader-Willi Syndrome
Edited by S. B. Cassidy

genes. Ten patients with AS and 7 with PWS were studied at the 700-850 band level using G- and R-band staining techniques on patients with PWS and, in addition, Q-band staining on the patients with AS. The chromosomes of the parents of the patients with AS, when available, were similarly studied and compared with those from their affected offspring. In all patients studied, there was a deletion present in the proximal long arm of chromosome 15. The deletions appeared variable in size, with the deletions in the patients with AS generally somewhat larger than those in the patients with PWS. The deletions were overlapping in location between the two syndromes, and in all cases part of all of band q11.2 was lost.

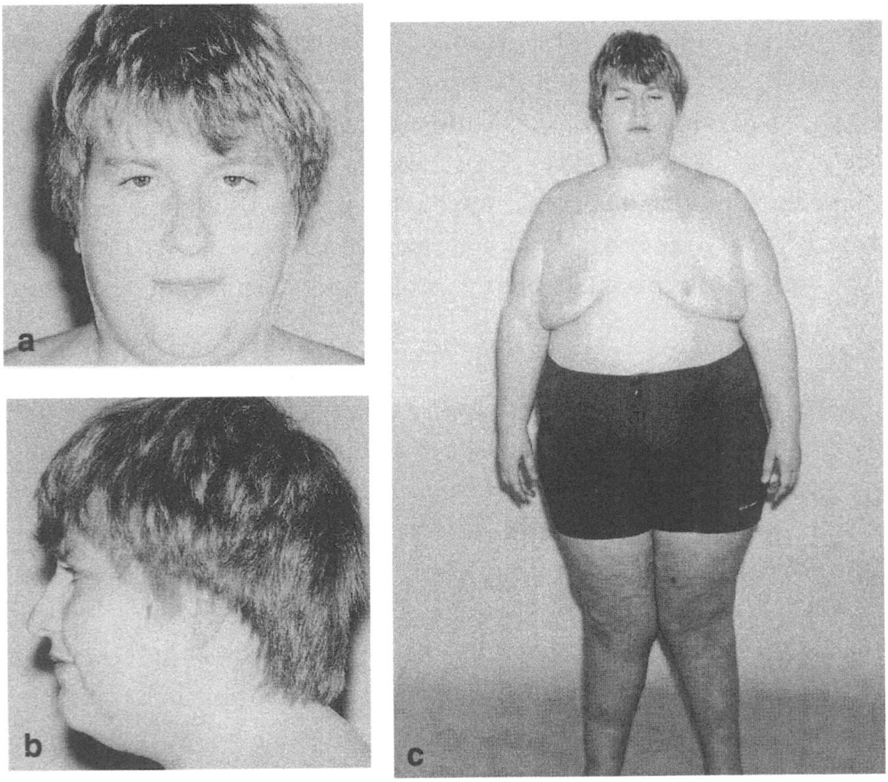

Figure 1. Twenty-four year old male (LH) with PWS and a deletion of chromosome 15(q11.2q12). a. Note blond hair, almond-shaped upslanted palpebral fissures, narrow bifrontal diameter, thin down-turned upper lip. b. Note short nose and mildly small mandible. c. Truncal obesity is evident.

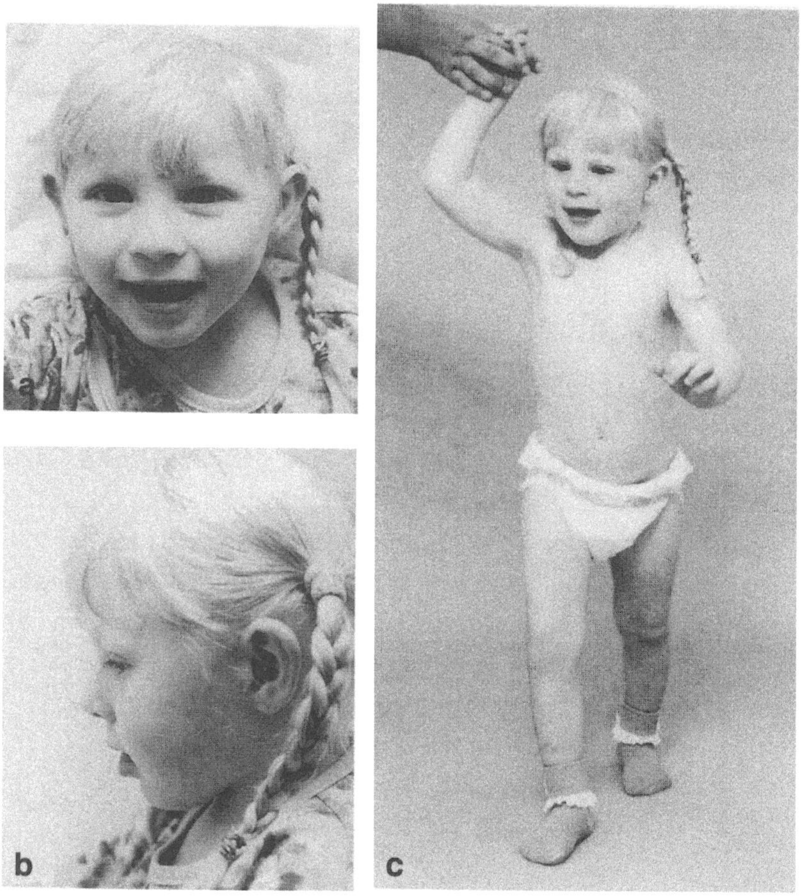

Figure 2. A 3 1/2 year old child with AS (AD). a. Note happy alert expression and blond hair. Few dysmorphic features are seen. b. Profile of same patient showing characteristic nose and mandible. c. Full body illustrating typical posture when walking.

Knoll et al (1989) studied the DNA from four patients with AS with five chromosome 15(q11.2-q13) specific cloned segments and compared results with their parents. They found no differences in the deletions between the patients with AS and PWS, but the parental origin differed. A maternal origin of the deletion in the patients with AS was demonstrated by use of restriction fragment length polymorphisms (RFLPs) in contrast to the paternal origin reported for the deletions in patients with PWS (Butler and Palmer et al, 1983; Butler et al, 1986; Mattei et al, 1983; Nicholls et al, 1989).

Our previous cytogenetic results (Magenis et al, 1990) in nine families supported the proposed exclusive maternal origin of the deletions of 15q in AS. In this paper, we extend the cytogenetic study of patients with AS and their families to 20 and the patients with PWS to 11.

METHODS

Late prophase - early metaphase chromosomes were prepared from peripheral blood lymphocytes from patients and parents (Yunis, 1976; Francke and Oliver, 1978). The preparations were stained with trypsin-Giemsa (Seabright, 1971) for high resolution G-bands and with chromomycin A_3/distamycin (Schweizer, 1980 modified by Magenis et al, 1988) for R-bands to delineate breakpoints. Q-banding with quinacrine mustard (Caspersson et al, 1970) and fluorescent R-banding were used to determine parental origin (Olson et al, 1986; Olson and Magenis, 1988).

RESULTS

The chromosomes 15 of all patients were at the 700-850 band stage with both G- and fluorescence R-bands. Judgements were made independently by at least two cytogeneticists for the presence and breakpoints of 15q cytogenetic deletions. The q11-13 region of chromosome 15 is difficult to assess; the region frequently stains poorly, is obscured by bends and curves, and the amount of heterochromatin extending from the centromere into the long arm is variable and not always quantifiable. In addition to these technical problems, there are discrepancies between the bands obtained in the high resolution material studied here and the bands in the idiograms in ISCN shown in Figure 3 (Harnden and Klinger, 1985). Band q12, noted in our material at the 550 band stage, splits into two gray bands with a pale band in between (q12.1; q12.2; q12.3) at the 850 band stage. In some of our patients, one of the breakpoints appears to be within q12, probably q12.2; to accommodate this, the ISCN

1985 idiogram of the proximal 15q region has been altered. As an example of these findings, the R- and G-banded chromosomes 15 from two cells from a patient with a moderately large deletion are illustrated (Figure 4); the chromosomes are aligned with an altered idiogram. Band q13.2 is illustrated as ISCN at the 850 band level though it does not appear until band levels greater than 850 (Figure 3, 4).

Eighteen of the 20 patients with AS had an interstitial deletion of the proximal long arm of chromosome 15. The G- and R-bands of six of these are shown in Figure 5.

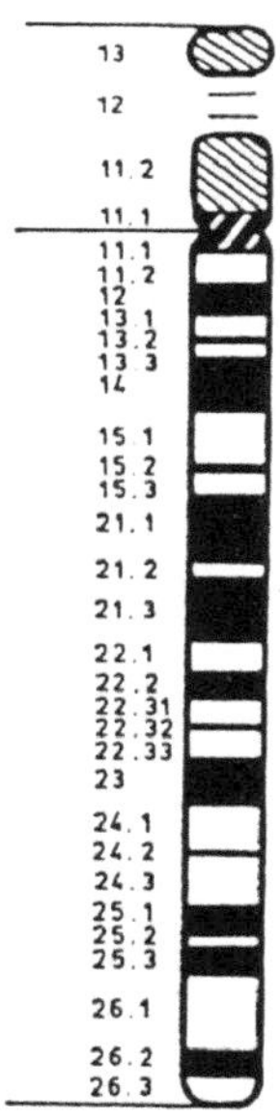

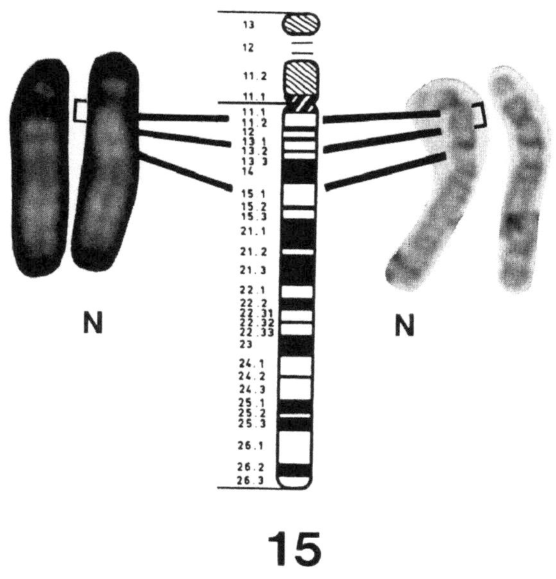

Figure 3. (left) Idiogram of band distribution on chromosome 15 at the 850 band level from ISCN 1985 (Harnden and Klinger, 1985).

Figure 4. (right) Idiogram of chromosome 15 at the 850 band level (Harnden and Klinger, 1985) modified in the q11.2-q13 region. Two pairs of chromosomes 15 at the 850 band stage from patient KS, one stained with fluorescent R-banding (left) and one with high-resolution G-banding (right), are illustrated. The abnormal chromosomes are on the far left and far right. Lines indicate bands present on the normal chromosomes, and brackets on the normal chromosomes show the deletion breakpoints. Note that band q12 in KS at this stage splits to form two smaller gray sub-bands separated by a pale band. This differs from ISCN but has been a consistent finding in our material.

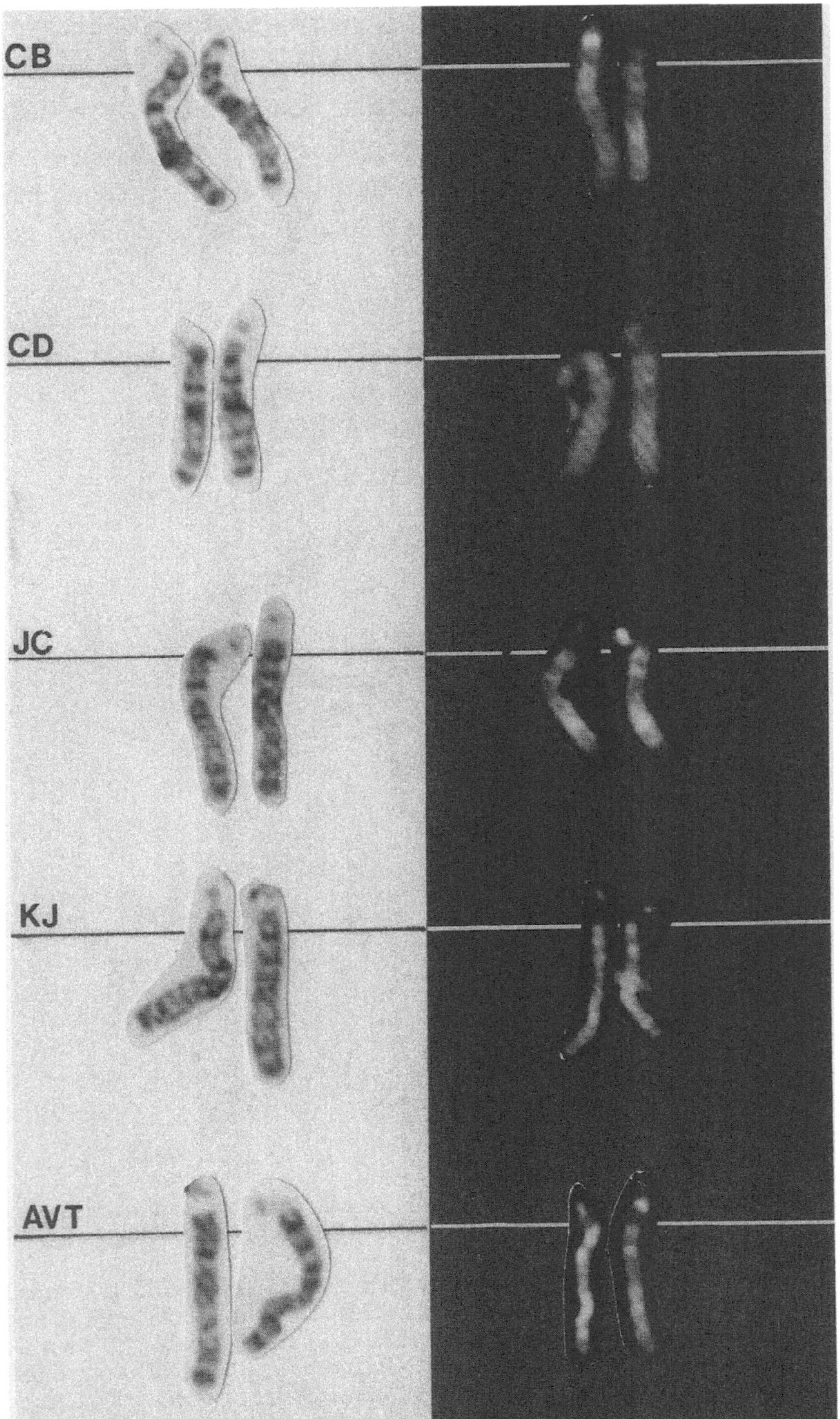

Figure 5. Pairs of G-banded (white background) and R-banded (black background) chromosomes 15 from five patients with AS. The deleted chromosomes are placed on the right in each pair.

The extent of the deletion in each of the 18 patients with AS is illustrated as a vertical bar on an altered 850 band stage idiogram in Figure 6. The deletions are variable in size, the largest in patient KS (bands q11.2-q13.2) and the smallest in RM (q11.2-q11.2)

At least one parent of 16 of the 18 patients was available for study. All parents studied had normal chromosomes; with Q- and R-band techniques the parental origin of the deleted chromosomes 15 in all 16 was determined to be maternal. The results of the parental origin studies in 3 of the 16 patients are shown in Figure 7.

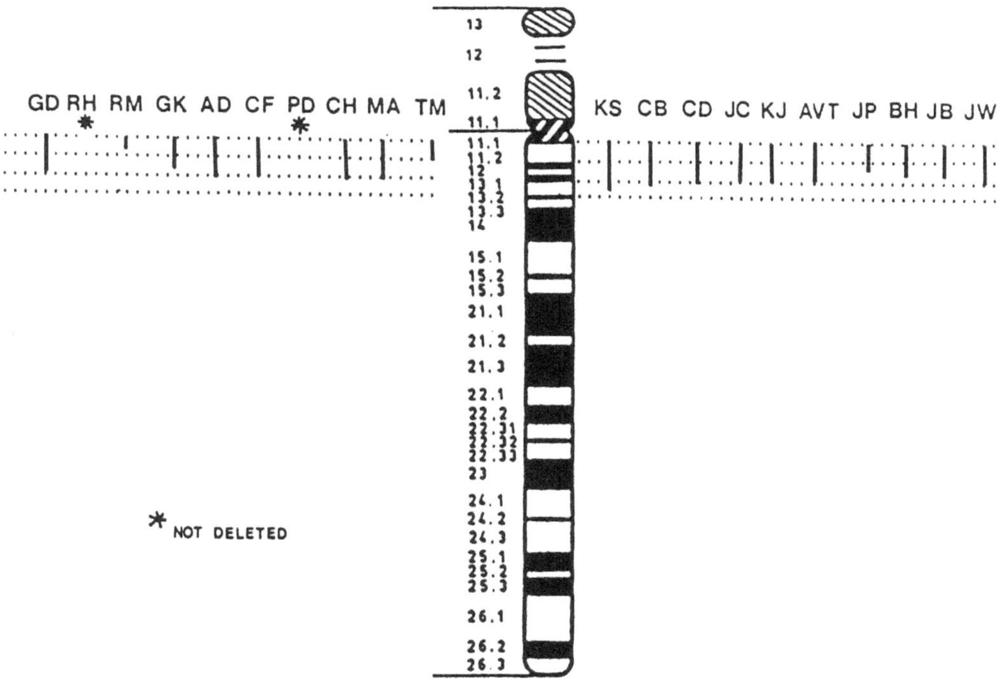

Figure 6. Summary of chromosome findings in the 20 patients with AS. The extent of the deletion in each patient, when present, is indicated on the diagram as a bar. Note that the deletions are variable in size but include a common region in q11.2.

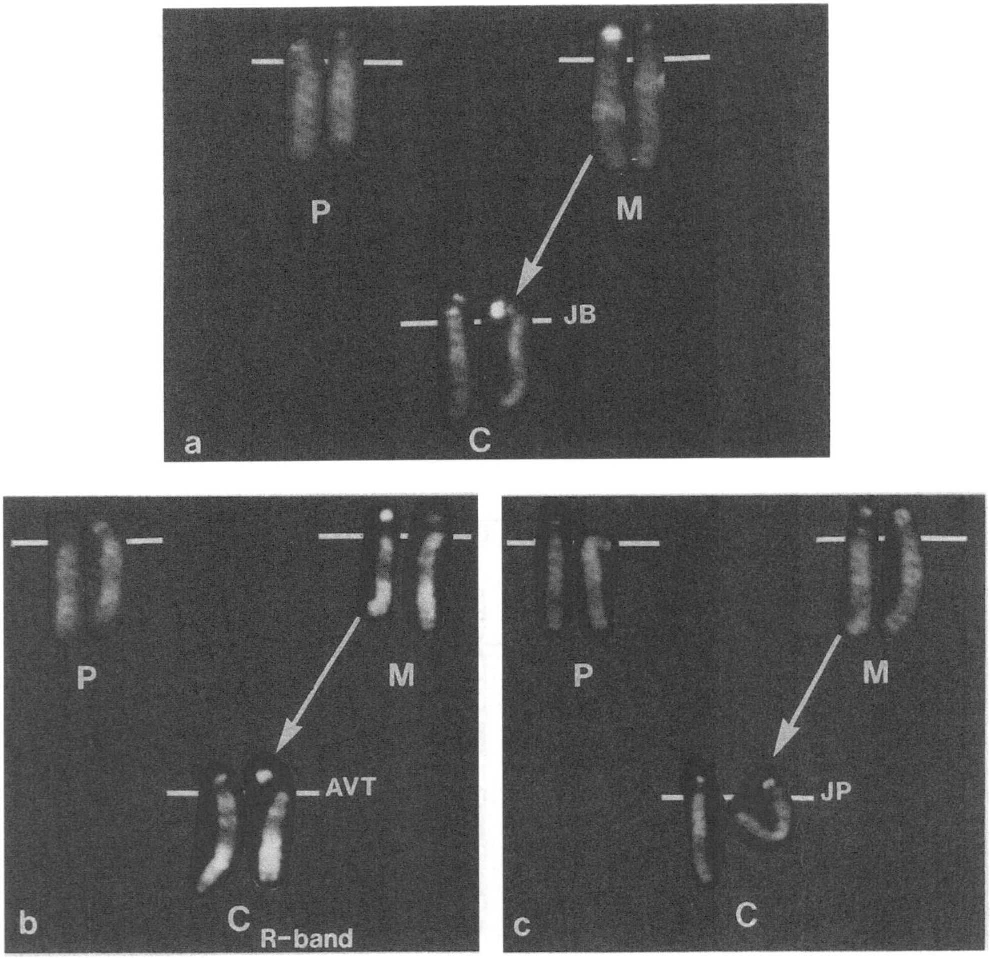

Figure 7. a-c. Quinacrine stained chromosomes 15 from patients JB and JP and parents, and R-banded chromosomes from patient AVT and parents. The deleted 15 from each patient is placed on the right. Arrows indicate the origin of the deleted 15 as shown by the similar fluorescent heteromorphisms. In each case, it was inherited from the mother.

Parental origin studies were also performed on the chromosomes of the two patients without detectable deletions to determine the presence or absence of uniparental disomy. In both cases, one of the chromosomes 15 in the patients could be traced to the mother, and the other to the father, providing no evidence for an irregular inheritance (Figure 8).

The chromosomes of 11 patients with typical PWS were studied with G- and R-banding techniques at 700-850 band levels. Nine of the 11 patients had a deletion of the proximal long arm of chromosome 15 similar to those found in the patients with AS (Figure 9).

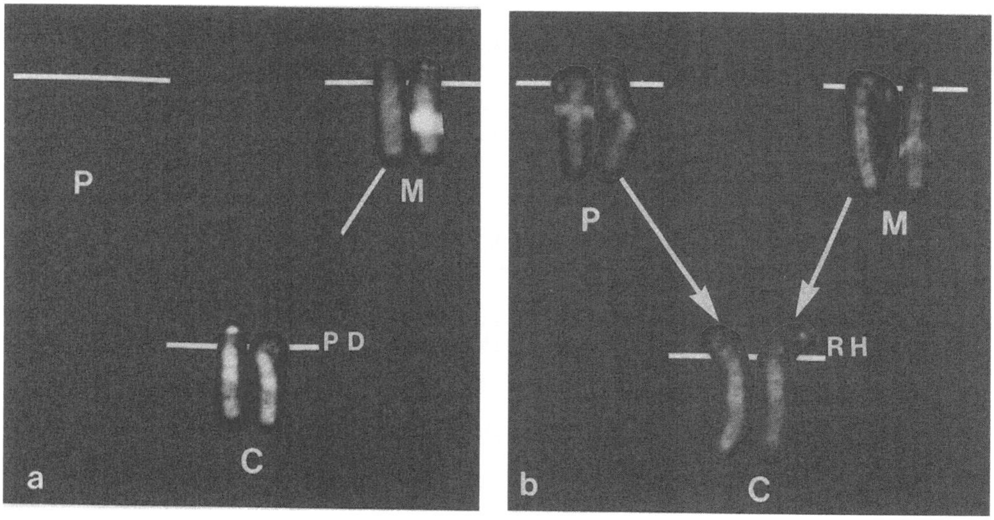

Figure 8a,b. Comparison of quinacrine variants for parental origin of the chromosomes 15 from the two patients without deletions. The father of PD in a. was not available for study. However, the quinacrine variant of one chromosome 15 in PD was identical to one of his mother's; the other chromosome 15 did not match either of the mother's 15's and thus must be from the father. It is possible but unlikely that the father would have a chromosome 15 variant identical to the one present in mother and child; thus uniparental disomy is also unlikely.

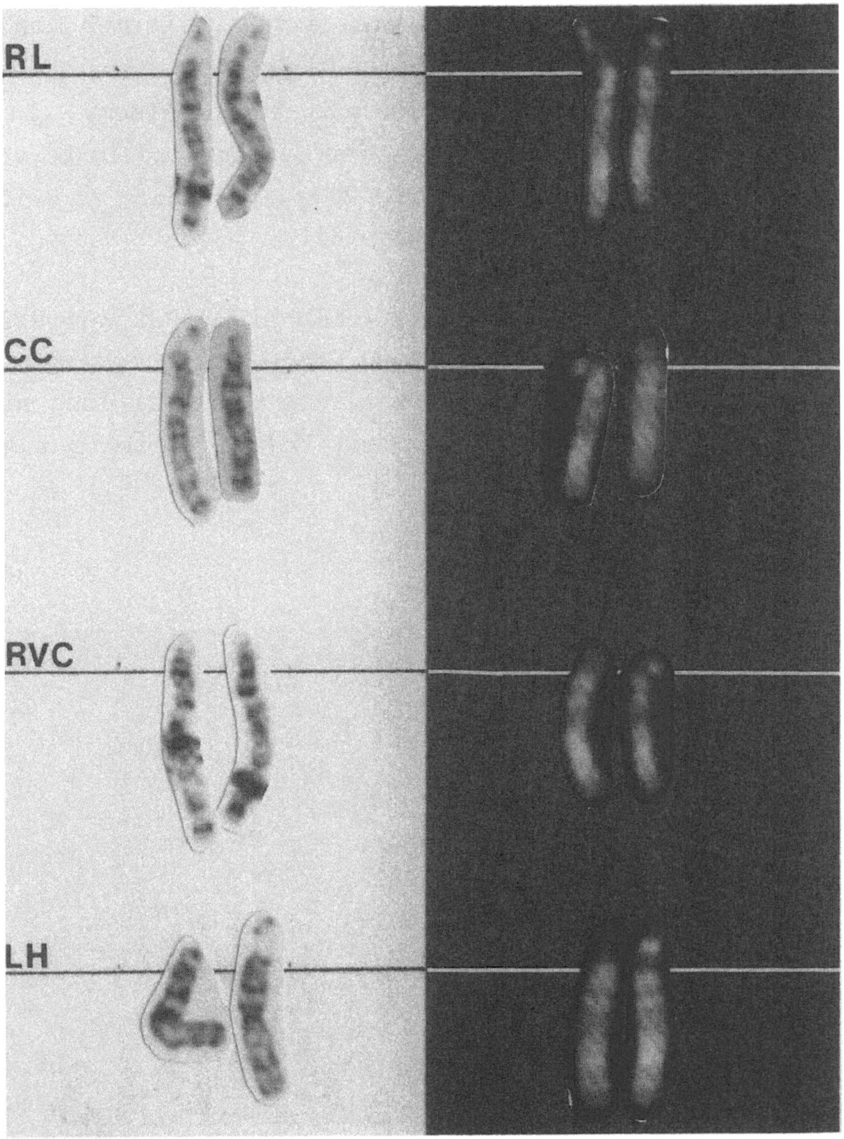

Figure 9. Pairs of G-banded (white background) and R- banded (black background) chromosomes 15 from four patients with PWS. The abnormal deleted chromosome is placed on the right of each pair.

The chromosomes of the other two patients did not reveal a deletion. Parental origin studies have not yet been done.

The size of the deletions observed in the patients with PWS studied is noted on the bar graph in Figure 10. The deletions are variable in length but in general smaller than the deletions seen in AS. In all cases, part of band 15q11.2 is missing.

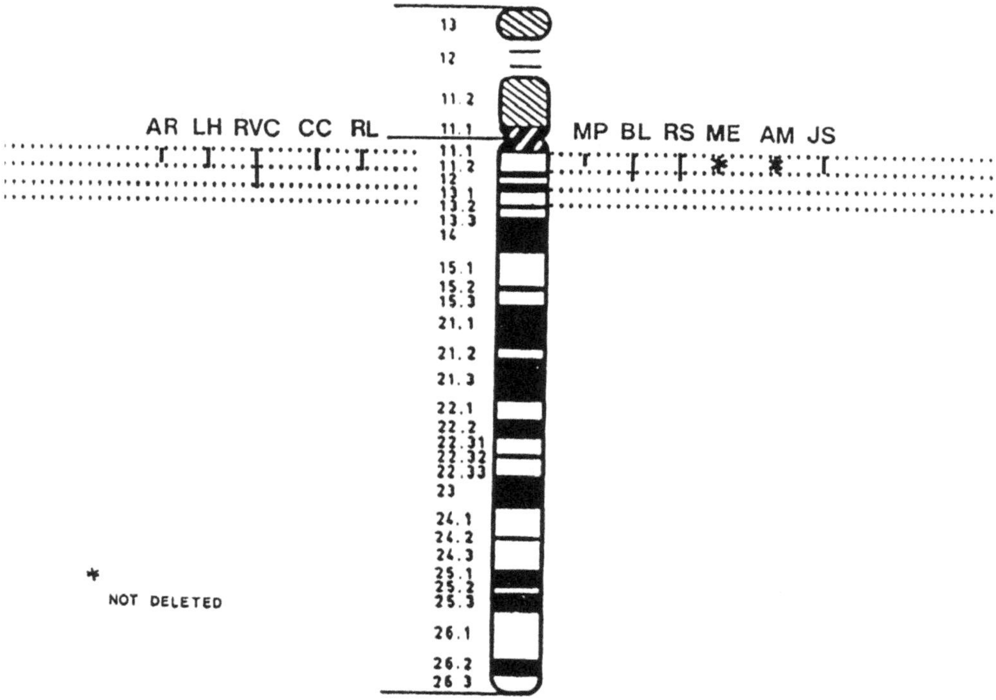

Figure 10. Summary of the chromosome results in the 11 patients with PWS. The extent of the deletion in each patient is indicated by a vertical bar, against an 850 band idiogram modified from ISCN 1985 (Harnden and Klinger, 1985). Note differing size deletions among the patients, but with the region q11.2 in common.

DISCUSSION

The incidence of the AS is not yet clear; it is clear that the condition is more frequent than was previously realized and has been underrecognized. The clinical phenotype is distinct and has minimal overlap with the phenotype of PWS. Overlap has been found in the location of the deletions of chromosome 15 in our patients, the common region of both conditions being 15q11.2. The importance of the q11.2 region to the phenotype is further supported by the fact that the size of the deletion seen cytogenetically has not correlated with the severity or with any specific manifestations noted clinically. Patient KS, for example, has a large deletion and is very similar in facial appearance and ability to JP, who has a smaller deletion. Thus, the postulate that the different phenotypes are a result of deletion of different segments of 15q appears unlikely. However, the confirmed differences in parental origin, preferentially paternal for PWS (Butler and Palmer, 1983; Butler et al, 1986; Mattei et al, 1983; Nicholls et al, 1989) and maternal for AS (Magenis et al, 1990; Donlon, 1988; Knoll et al, 1989; Pembrey et al, 1989; Cooke et al, 1989; Williams et al, 1989b) must be a factor in the expressed phenotypes.

In mice it has been shown experimentally that there is a differential effect of uniparental disomy for certain chromosomes and certain chromosome regions depending on whether the disomic chromosomes or chromosome segment comes from the male or the female parent (Cattanach and Kirk, 1985; Cattanach, 1986). Mice carrying translocations were chosen such that each parent carried the same rearrangement that was known to undergo a high rate of nondisjunction. With the occurrence of the same type of abnormal disjunction in both parents and union of complementary products, young were produced that were chromosomally balanced but that had either maternal duplication/paternal deficiency or paternal duplication/maternal deficiency for certain regions of the involved chromosomes. With some of these chromosomes, phenotypic differences dependent on the parent involved could be detected at birth. Maternal duplication/paternal deficiency for distal chromosome 2 results in a flat-sided, arch-backed hypokinetic newborn mouse that has ineffective suck and dies within 24 hours. The reciprocal type, paternal duplication/maternal deficiency, has short, square bodies, broad and flat backs, and are notably hyperkinetic. This type usually survives several days, but they develop tremor and balance problems. Similarity to the PWS/AS phenotypes is evident.

Solter (1988) recently reviewed the evidence for differential imprinting and expression of maternal and paternal genomes that utilizes manipulative techniques of nuclear transfer and genetic analysis of meiotic nondisjunction. This differential expression is shown by both methods to be necessary for normal development. The differential expression has been postulated to be due to differential methylation.

Chaillet et al (1991) assessed changes in the DNA methylation of an imprinted transgene as it switched from one parental sex to the other while traveling through gametogenesis and embryogenesis. They found that the maternally and paternally inherited methylation patterns are erased in the primodial germ cells and emerge during germ cell maturation.

It follows, then, that the different conditions, AS and PWS, may be the result of lack of the maternal component (AS) versus lack of the paternal component (PWS) in the same critical segment of this apparently unstable region of chromosome 15, rather than a dose effect of specific and different neighboring structural genes.

SUMMARY

Similar deletions of chromosome 15 long arm have been shown in both the Prader-Willi and Angelman syndromes. The distinctiveness of the phenotypes however, belies identical etiology. To investigate possible differences in deletion size and/or location, the chromosomes of 20 patients with AS and their parents, as well as the chromosomes of 11 patients with PWS, were studied with the use of three cytogenetic techniques, and the results compared. Giemsa-banding and fluorescence R-banding were used in all patient studies for assessment of size and location of the deletion; Q-banding and fluorescence R-banding were used for comparison of cytogenetic variants for parental origin.

Eighteen of the 20 patients with AS had deletions, with the proximal break in q11.2 and the distal break in q11.2, q12 or q13. Two patients had no visible deletion, but in each case one chromosome 15 was from the mother, the other from the father, with no evidence for uniparental disomy. In all 16 patients with AS with a deletion for whom parents were available, the deletion was maternally derived.

The results of cytogenetic deletion analysis of the patients with PWS were similar. The deletions appeared in general to be smaller than those found in AS with somewhat different distal breakpoints, but all included part of band q11.2. It is known from earlier studies that most, if not all, deletions in PWS are of paternal origin.

Differential expression of maternally- and paternally-derived genomes has been known and accepted in mice and other organisms. It now appears that similar differential expression, at least for some regions of the genome, occurs in the human as well.

REFERENCES

Angelman J (1965) "Puppet" children: A report on three cases. Dev Med Child Neurol 7:681-683

Butler MG, Palmer CG (1983) Parental origin of chromosome 15 deletion in Prader-Willi syndrome. Lancet 1:1285-1286

Butler MG, Meaney FJ, Palmer CG (1986) Clinical and cytogenetic survey of 39 individuals with Prader-Labhart-Willi syndrome. Am J Med Genet 23:793-809

Caspersson T, Zech L, Johnasson C, Modest EJ (1970) Identification of human chromosomes by DNA-binding fluorescent agents. Chromosoma 30:215-217

Cattanach BM, Kirk M (1985) Differential activity of maternally and paternally derived chromosome regions in mice. Nature 315:496-498

Cattanach BM (1986) Parental origin effects in mice. J Embryol Exp Morph Supp 97:137-150

Chaillet JR, Vogt TF, Beier DR, Leder P (1991) Parental-specific methylation of an imprinted transgene is established during gametogenesis and progressively changes during embryogenesis. Cell 66:77-83

Cooke A, Tolmie JL, Glencross FJ, Boyd E, Clarke MM, Day R, Stephenson JBP, Connor JM (1989) Detection of a 15q deletion in a child with Angelman syndrome by cytogenetic analysis and flow cytometry. Am J Med Genet 32:545-549

Donlon TA (1988) Similar molecular deletions on chromosome 15q11.2 are encountered in both the Prader-Willi and Angelman syndromes. Hum Genet 80:322-328

Francke U, Oliver N (1978) Quantitative analysis of high-resolution trypsin-Giemsa bands on human prometaphase chromosomes. Hum Genet 45:137-165

Harnden DG, Klinger HP (eds) (1985) An International System for Human Cytogenetic Nomenclature (1985). S Karger, New York

Imaizumi K. Takada F, Kuroki Y, Maritomi K, Hamabe J. Niikawa N (1990) Cytogenetic and molecular study of Angelman syndrome. Am J Med Genet 35:314-318

Kaplan LC, Wharton R, Elias E, Mandell F. Donlon T, Latt SA (1987) Clinical heterogeneity associated with deletions in the long arm of chromosome 15: Report of 3 new cases and their possible genetic significance. Am J Med Genet 28:45-53

Knoll JHM, Nicholls RD, Magenis RE, Graham JM Jr, Lalande M, Latt SA (1989) Angelman and Prader-Willi syndromes share a common chromosome deletion but differ in parental origin of the deletion. Am J Med Genet 32:285-290

Ledbetter DH, Riccardi VM, Airhart SD, Keenan BS, Crawford JD (1981) Deletions of chromosome 15 as a cause of the Prader-Willi syndrome. N Engl J Med 304:325-329

Magenis RE, Brown MG, Lacy DA, Budden S (1987) Is Angelman syndrome an alternate result of del(15)(q11-q13)? Am J Med Genet 28:829-838

Magenis RE, Toth-Fejel S, Allen LJ, Black M, Brown MG, Budden S, Cohen R, Friedman JM, Kalousek D, Zonana J, Lacy D, LaFranchi S, Lahr M, MacFarlane J, Williams CPS (1990) Comparison of the 15q deletions in Prader-Willi and Angelman syndromes: Specific regions, extent of deletions, parental origin and clinical consequences. Am J Med Genet 35:333-349

Magenis RE, Sheehy RR, Brown MG, McDermid HE, White BN, Zonana J, Weleber R (1988) Parental origin of the extra chromosome in the cat eye syndrome: Evidence from heteromorphism and in situ hybridization analysis. Am J Med Genet 29:9-19

Mattei JF, Mattei MG, Girand F (1983) Prader-Willi syndrome and chromosome 15: A clinical discussion of 20 cases. Hum Genet 64:356-361

Nicholls RD, Knoll JH, Glatt K, Hersh JH, Brewster TD, Graham JM Jr, Wurster-Hill D, Wharton R, Latt SA (1989) Restriction fragment length polymorphism within proximal 15q and their use in molecular cytogenetics and the Prader-Willi syndrome. Am J Med Genet 36:66-77

Olson SB, Magenis RE, Lovrien EW (1986) Human chromosome variation: The discriminatory power of Q-band heteromorphism (variant) analysis in distinguishing between individuals with specific application to cases of questionable paternity. Am J Hum Genet 38:235-252

Olson SB, Magenis RE (1988) Preferential paternal origin of de novo structural rearrangements. In Daniel A (ed) The Cytogenetics of Mammalian Autosomal Rearrangements. pp 583-599, Alan R. Liss, Inc, New York

Pembrey M, Fennel SJ, Van den Berghe J, Fitchett M, Summer SD, Butler L, Clarke C, Griffiths M, Thompson E, Super M, Baraitser M (1989) The association of Angelman's syndrome with deletions within 15q11-13. J Med Genet 26:73-77

Prader A, Labhart A, Willi H (1956) Ein Syndrome von Adipositas, Kleinwuchs, Kryptorchidism und Oligophrenie nach myatonicartigem Zustand in Neugeboren alter. Schweiz Med Woschenschr 86:1260-1261

Schweizer D (1980) Simultaneous fluorescent staining of R-bands and specific heterochromatic regions (DA-DAPI bands) in human chromosomes. Cytogenet Cell Genet 27:190-193

Seabright M (1971) A rapid banding technique for human chromosomes.Lancet 2:971-972

Solter D (1988) Differential imprinting and expression of maternal and paternal genomes. Annu Rev Genet 22:127-146

Williams CA, Hendrickson JE, Cantu ES, Donlon TA (1989a) Angelman syndrome in a daughter with del(15)(q11q13) associated with brachycephaly, hearing loss, enlarged foramen magnum and ataxia in the mother. Am J Med Genet 32:333-338

Williams CA, Gray BA, Hendrickson JE, Stone JW, Cantu ES (1989b) Incidence of 15q deletions in the Angelman syndrome: A survey of 12 affected persons. Am J Med Genet 32:339-345

Yunis JJ (1976) High resolution of human chromosomes. Science 191:1268-1270

MOSAICISM FOR DELETION 15q11q13 IN SPORADIC AND FAMILIAL CASES

Shivanand R. Patil
Division of Medical Genetics
Department of Pediatrics
University of Iowa Hospitals and Clinics
Iowa City, Iowa
USA

Bryce Hauschildt[1], Marie Greally[1], Chris Headley[1], Denise Wilson[1], James Hanson[1] and Timothy Donlon[2]

Department of Pediatrics, Division of Medical Genetics, University of Iowa Hospitals and Clinics, Iowa City, Iowa[1]; Department of Pathology, Stanford University Hospital, Stanford, California[2], USA

INTODUCTION

An interstitial deletion of 15(q11q13) has been well documented in some patients with Prader-Willi syndrome (PWS) and Angelman syndrome (AS) (Ledbetter et al, 1981; Kaplan et al, 1987; Magenis et al, 1987). Although a majority of the PWS or AS cases occur sporadically, a few families with two or more relatives have been reported (Burke et al, 1987; Lubinsky et al, 1987; Baraitser et al, 1987; Willems et al, 1987). No deletion was detected in these familial cases, suggesting either an autosomal recessive mode of inheritance or a new autosomal dominant mutation.

A number of sporadic cases have also been described with deletion 15(q11q13) but without the features of PWS or AS (Schwartz et al, 1985; Kaplan et al, 1987; Reynolds et al, 1987; Greenberg and Ledbetter, 1987).

However, the existence of mosaicism for deletion 15(q11q13) has not been documented since the report by Cassidy et al (1984), and it may have been overlooked because of interpretative problems. They reported that two out of twelve cases studied were mosaic for the deletion. We report here cytogenetic and limited molecular data on three sporadic cases of mosaicism and on two families where

NATO ASI Series, Vol. H 61
Prader-Willi Syndrome
Edited by S. B. Cassidy

the mothers were mosaic and their children were non-mosaic for the deletion.

MATERIALS AND METHODS

Lymphocytes from peripheral blood samples on all patients were stimulated with phytohemagglutinin, cultured for 96 hours and synchronized to obtain elongated chromosomes. Fifteen to 20 G-banded cells were analyzed at 550 or greater band length. Whenever possible, a repeat blood sample was obtained to confirm the initial chromosome findings.

Molecular studies were carried out on a limited number of our patients. DNA was extracted from peripheral blood samples (as well as from fibroblasts of patients 4, 5 and 6) by standard methods (Aldridge, 1984), digested with restriction enzymes under conditions recommended by the manufacturer (New England Biolabs), and size fractioned by agarose gel electrophoresis. The following cloned DNA segments (Donlon et al, 1986) were radiolabelled (Feinberg and Vogelstein, 1983) to high specific activity, hybridized to Southern blots (Southern, 1975) and exposed to Kodak X-OMAT AR film using DuPont Lightning Plus intensifying screens, at -80C: pTD189-1 (D15S13) was hybridized to TaqI digested samples, pTD3-21 (D15S10) was hybridized to TaqI digested samples, pIR10-1 (D15S12) was hybridized to ScaI digested samples, and pIT39 (D15S18) was hybridized to BglII digested samples.

CLINICAL HISTORIES

Sporadic Cases

Patient 1, JJ, was the 4300g product of a normal pregnancy and delivery. There was meconium aspiration at birth and Apgar scores were 5 and 6 at 1 and 5 minutes, respectively. Maternal and paternal ages were 20 and 25 years, respectively, at the time of his birth.

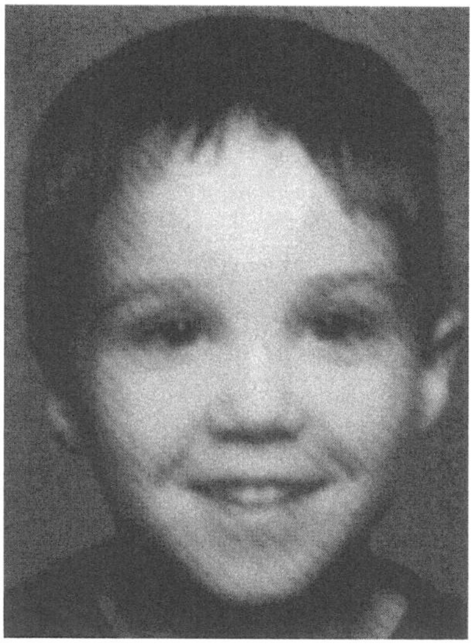

Figure 1a. Facial features of patient 1 at 44 months include: broad forehead, telecanthus, flat midface, flat and long philtrum and small mouth.

JJ was first evaluated at age 3 months because of dysmorphic facial features and increased birth weight (>95th centile). Early milestones were delayed. He did not walk until 18 months of age and speech development was delayed. An evaluation at 42 months indicated an approximate delay of 12 months in speech and language development. Fine motor skills were appropriate for age, but gross motor skills were at the 31 month level. Non-verbal intelligence tests were within the normal range.

On examination at 44 months, JJ was tall and slim with a height of 106cm (90th centile) and a weight of 15kg (25th centile). His head circumference was 50.5cm, which is at the 50th centile for his age. He had dysmorphic facial features (Figure 1a) which included flat mid-face, large forehead, broad nasal bridge, epicanthal folds, almond-shaped eyes and upturned nose. He had a narrow bitemporal diameter, long philtrum, small mouth, crowded teeth and simple ears. There was some speckling of the iris bilaterally. Hyperflexible finger joints were noted and bilateral clinodactyly of the fifth finger was present. At birth there was redundant skin at the nape of the neck, but this became less evident with age. He also had mild laryngomalacia. Mild hypotonia with normal reflexes was noted on neurological examination.

Family history was negative for spontaneous abortion, mental retardation or seizure disorder. There was maternal scoliosis, and the paternal grandmother was diabetic.

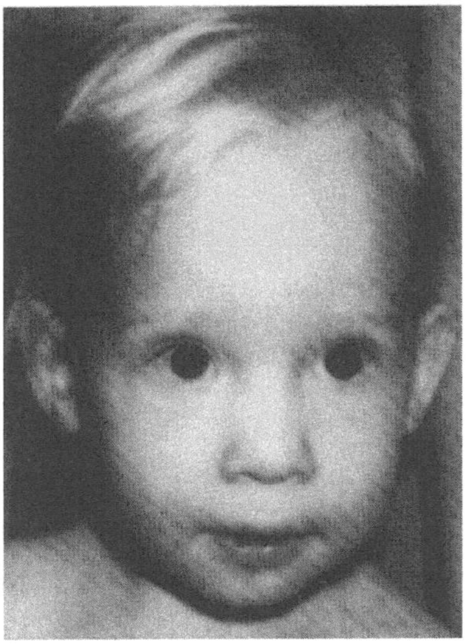

Figure 1b. Facial features of patient 2 at 16 months of age showing long narrow face, prominent forehead, broad nasal bridge, prominent premaxilla, long philtrum, bilateral frontal hair upsweep.

Patient 2, TP, was the third female child born to a healthy, non-consanguineous couple. Pregnancy was uncomplicated and delivery was normal. Apgar scores were not available, but were reportedly normal. Birth weight was 3000g and length was 19 inches. Maternal and paternal ages were 23 and 25 years, respectively, at the time of her birth.

At two weeks of age, a heart murmur was heard and a diagnosis of ventricular septal defect (VSD) was made. Problematic vomiting occurred from the second to third months but resolved spontaneously. At 16 months she was evaluated for developmental delay and failure to thrive. At that time she was crawling and could stand with support. She rolled over at one year. She had no speech and babbled frequently.

Neurological evaluation included MRI and head CT with contrast. The results revealed mild diffuse ventriculomegaly, equivocal cerebral atrophy, no masses, and smaller than expected brainstem size.

She was first seen for genetic evaluation at 16 months of age. At that time her height, weight and head circumference were all below the 5th centile. She walked at age 2 years. On review at 4 years and 1 month of age, her height was 94.6cm and weight was 12.4kg. Both values are below the 5th centile as was her OFC. She was a very cooperative child and responded well to questions. She did not speak until she was three years of age, but responded well to stimulation from a special education program. Her facial features (Figure 1b) included a long narrow face, prominent forehead, broad nasal bridge, mild telecanthus, high arched palate, prominent pre-maxilla with large overjet, long upper lip, mild micrognathia, and bilateral frontal hair upsweep. She had two hair whorls. Her hair was sparse, dry and brittle and she had a previous diagnosis of trichorrhexis nodosa. A mild bilateral valgus deformity of the halluces was noted. Her gait was broad-based and slightly unsteady.

Family history was negative for seizures and mental retardation. A paternal uncle had leukemia and died at five years of age. An older sibling also had trichorrhexis nodosa.

Patient 3, KH, was the 2500g product of a pregnancy which was complicated by an equivocal non-stress test. Labor was induced and lasted approximately 3 1/2 hours. Apgar scores were 2 and 2 at 1 and 5 minutes, respectively. The patient was placed on a ventilator for 12 hours and then easily weaned. At two days of age he had several episodes of brachycardia without apnea or desaturation. Head CT revealed a subarachnoid hemorrhage which eventually resolved.

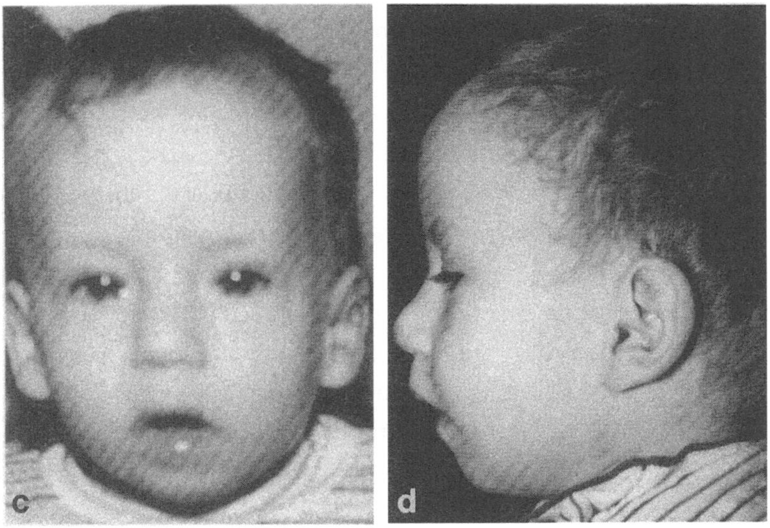

Figure 1c. Patient 3 at age 15 months with broad forehead, mild telecanthus and long upper lip. d. Side view of patient 3 showing prominent pre-maxilla and micrognathia.

At the time of KH's birth, his mother was 35 years of age and his father was 40. On physical examination at 14 months the patient's height was 66.3cm, he weighed 6.65kg and his head circumference was 44.5cm. All these values are well below the 5th centile. He had (Figure 1c and d) a prominent metopic suture, a patent anterior fontanelle measuring approximately 2.2cm x 2.2cm, slightly upslanted palpebral fissures, and pale blue irises with slight speckling. His outer canthal distance was 7.5cm (75th centile), inner canthal distance was 2.5cm (approximately 50th centile) and interpupillary distance was 4.5cm (<50th centile). His right ear showed mild posterior rotation with folding of the upper helix. He had a slightly beaked nose with a prominent pre-maxilla, long philtrum and micrognathia. He had a cleft of the soft palate and the posterior portion of the hard palate. Neurological examination revealed hypertonia. At 15 months he was trying to stand by himself. He could not crawl, was still unable to sit without support and had no speech. An eye examination at 17 months was normal except for mild hyperopia.

The parents of KH were healthy and unrelated. There is no family history of spontaneous abortion or seizures. The mother was described as a "slow learner and low functioning". She was obese and is 4 feet and 11 inches in height.

Familial Cases

Family 1:

Patient 4, AR, was the 3400g product of a normal pregnancy and delivery. He was the first child of healthy non-consanguineous parents. Maternal and paternal ages were 24 and 29 years, respectively, at the time of his birth. Throughout his first years of life, AR was small and demonstrated poor growth, with height and weight below the 5th centile. Otherwise, he was in excellent health with no reported feeding difficulties.

Figure 2a. Picture of two brothers in Family 1 (Patient 4, AR at age 48 months, on the left and patient 5, JR at 31 months, on the right). AR has a slightly bulging forehead, depressed and broad nasal bridge, hypertelorism, telecanthus and flat philtrum. JR has a prominent forehead, deep set eyes, broad nasal bridge, hypertelorism, telecanthus and small mouth.

The propositus was seen initially in a pediatric neurology clinic because of hypotonia, poor growth and developmental delay. On physical examination at 3 years 11 months, his height was less than the 5th centile (90cm) and his weight was at the 10th centile.

Head circumference was 50.5cm. He had mild dysmorphic facial features (Figure 2a) which included a slightly bulging forehead, hypertelorism and telecanthus. A right frontal hair upsweep was noted. Ears were normal in shape and position. Hand size was small (<5th centile) and fingers were tapering with short distal phalanges and short nails. A single transverse palmar crease was noted on the left hand. Foot size was also small (<5th centile) and there was skin syndactyly of the 2nd and 3rd toes of the right foot. Neurological examination revealed moderate hypotonia with normal reflexes. Joint hyperextensibility of hands and fingers was noted. Genitalia were normal.

Developmental milestones in AR were significantly delayed. He did not crawl until 18 months and walked at 30 months of age. He did not begin to use simple sentences until he was three years of age. Bone age x-rays which were carried out at chronological age of 33 months revealed a bone age of 14-19 months.

There was no history of spontaneous abortion or seizures in this family. The mother was placed in foster care as a baby and no details of her family were available. She had a learning disability which required special education classes throughout her schooling.

Patient 5, JR, is the younger brother of AR, born after a normal pregnancy and delivery. He is 17 months younger than AR. He was placed in foster care at two years of age because of poor growth and failure to thrive but otherwise was in good health with no reported neonatal feeding difficulties.

On physical examination at 2 years 7 months, JR was a pleasant, alert and cooperative child. He was very small with a height of 82.8cm (<5th centile). His weight was 11.6kg (5th centile) and his head circumference was 47.2cm (<5th centile). He had mildly dysmorphic facial features (Figure 2a) including a prominent forehead, deep-set eyes, broad nasal bridge, hypertelorism, telecanthus and a small mouth with thin upper lip. His hand size was less than the 5th centile, with skin dimpling over the knuckles, tapering fingers and normal nails. He had bilateral clinodactyly of the fifth digits. JR's foot measurements were at the 10th centile. Genitalia were normal. Neurological examination revealed mild hypotonia with normal reflexes.

Early developmental milestones were normal. He crawled at 7 to 8 months but did not walk until 18 months. Speech development was normal.

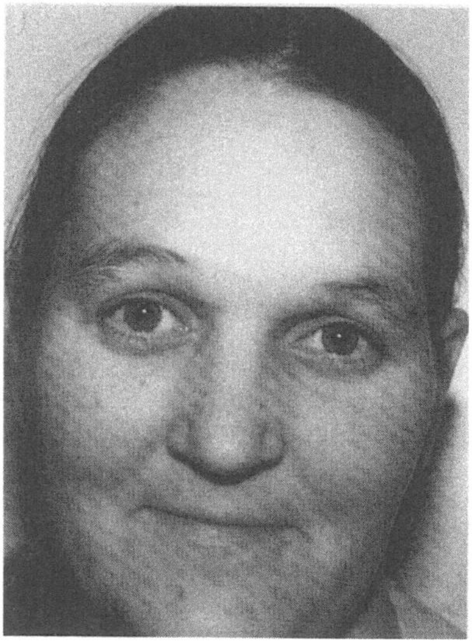

Figure 2b. Mother (28 years) of patients 4 and 5 showing broad forehead.

Patient 6, AMR, is the 28 year old mother of AR and JR. On physical examination her height was five feet four inches. Her hands were small with short tapering fingers, short distal phalanges, short nails and hyperextensible finger joints. She had bilateral skin syndactyly of the second and third toes. Facial features (Figure 2b) were unremarkable apart from a large forehead.

Family 2:

Patient 7, SL, was the 3500g product of a normal pregnancy and delivery. His parents were healthy and unrelated. Maternal and paternal ages were 21 and 23 years, respectively, at the time of his birth.

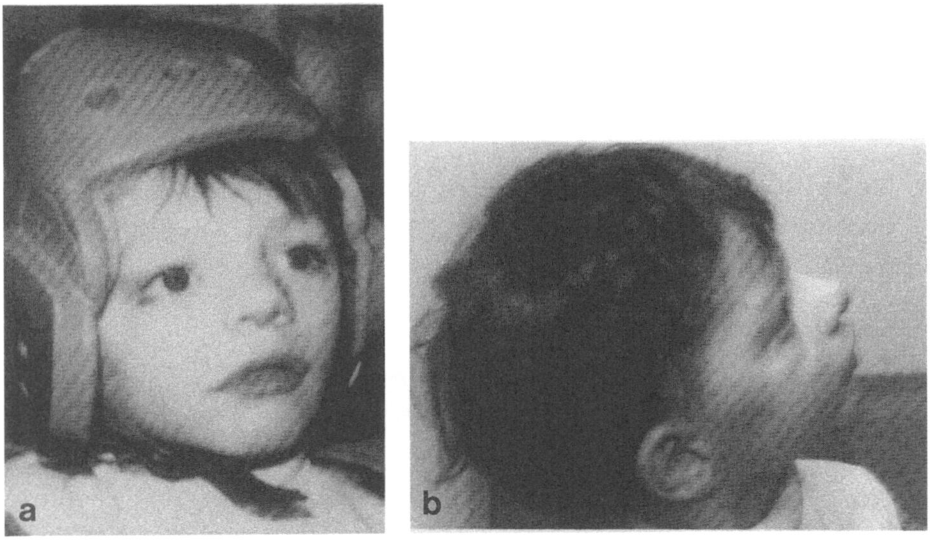

Figure 3a. Picture of patient 7 at 31 months showing right facial hypoplasia, flattened right malar region, bulbous nose, downslanting palpebral fissures, and macrostomia. Protective helmet was worn because of self-destructive behavior. b. Sideview picture of patient 7 showing small forehead, prominent nose, prominent pre-maxilla and flattened right malar region.

On examination at 2 years 7 months, his height was 100cm (50th centile), weight was 14kg (10th-25th centile) and head circumference was 45cm (5th centile). He had plagiocepahly with flattening of the left occiput. His facial features (Figure 3a) included an asymmetric face with apparent right-sided hypoplasia, flattened right malleolar region, prominent bulbous nose, downslanting palpebral fissures, long eyelashes, hypertelorism and telecanthus. Ears were normal. Slight broadening of all fingertips was noted. Hand size was normal at 11.5cm. His feet were small, measuring 14cm (3rd centile). Genitalia were normal. Bilateral inguinal hernia repair was done at 4 months of age.

Neurological examination revealed a severely delayed child with no speech and with continuous self-stimulatory behavior including head banging, head slapping and biting of hands. Muscle bulk was noted to be decreased as was muscle tone.

Patient 8, PL, is the 25 year old mother of SL. She had a borderline personality disorder, mild mental retardation and mild microcephaly. Her head circumference was 54cm (20th centile). It was not possible to carry out a physical examination of this patient.

The first child of PL is a 4 year 10 month old girl, BL (patient 9). On evaluation for developmental delay, she was noted to have bilateral pes planus, clinodactyly of the 5th toes bilaterally, lax elbow and knee joints, but no dysmorphic features. She showed mild delay in motor and speech development. She was very tall with a height of 114cm (>95th centile) and a weight of 16kg (50th centile). She had an OFC of 47.5cm (2nd centile).

RESULTS

Cytogenetic Studies

Deletion in the 15(q11q13) region was seen in 50% to 80% of the cells analyzed in the sporadic cases (Figure 4). In two patients (1 and 3) it was possible to obtain a repeat sample, and the existence of the mosaicism was confirmed. Chromosome studies were carried out in both parents of patients 1 and 2 and in the mother of patient 3. Chromosome 15s in all these individuals appeared to be normal.

In the familial cases, the mother of patients 4 and 5 had the deletion in nearly 50% of cells, and the father of these patients was found to be chromosomally normal. Both patients had the deletion in all analyzable cells (Figure 5). Similarly, in family 2, the mother had the deletion in more than 50% of the analyzable cells and her son had the deletion in all cells (Figure 6).

Molecular Studies

Four DNA segments, which are known to be localized to 15(q11.2-12) and are also known to be deleted in a majority of PWS patients who have a recognizable cytologic deletion, were used as probes on DNA from blood and/or fibroblasts.

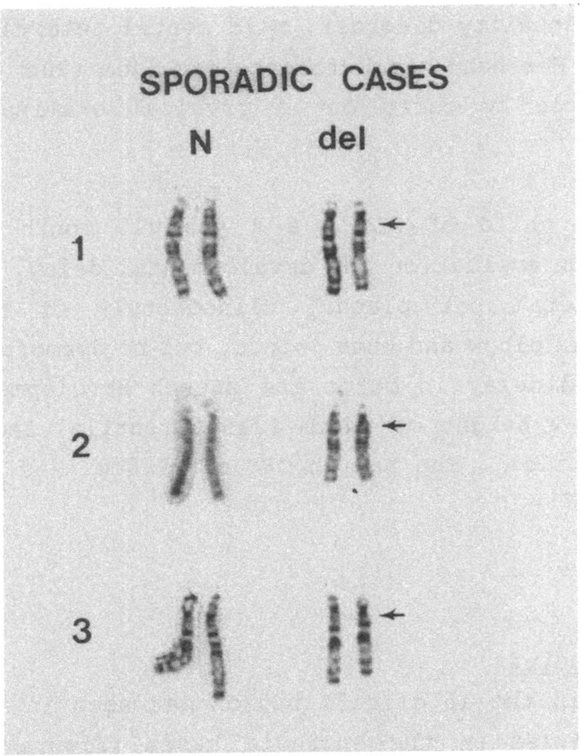

Figure 4. Partial kayotypes of chromosome 15s in sporadic cases. Chromosome 15 with the deletion is to the right side (arrow) of its normal homolog.

In patient 1 (Figure 7), the pIR10-1 probe demonstrated homozygosity for a 12.5 Kb allele in the father, and heterozygosity for a 12.5 Kb and 16.0 Kb allele in the mother. The patient was hemizygous for the 16.0 Kb allele, and he was apparently missing the 12.5 Kb allele from the father.

Molecular studies with the probes utilized were uninformative in patients 2 and 3.

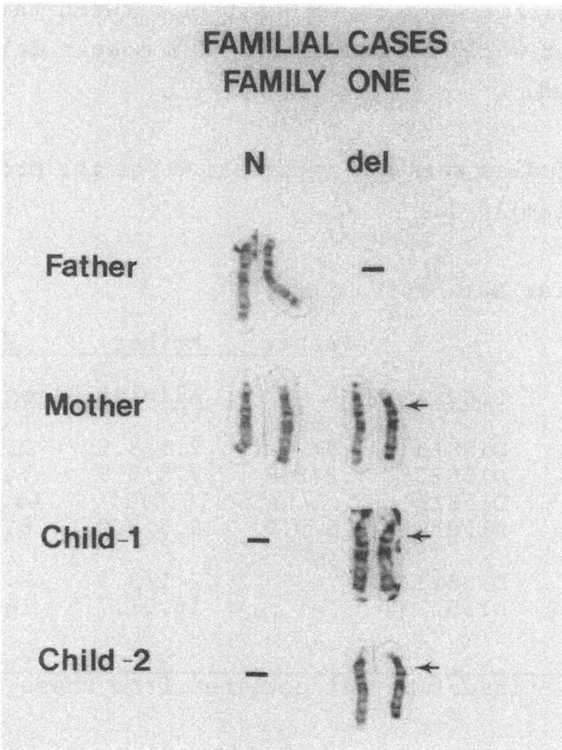

Figure 5. Partial karyotypes of chromosome 15s in members of family 1. Chromosome 15 with the deletion is to the right side (arrow) of its normal homolog.

In family 1 (Table I), patients 4 and 6 were heterozygous for locus D15S10, identified by probe pTD3-21; all contained 7.8 and 8.9 Kb alleles. Similarly, locus D15S12, identified by probe pIR10-1, was also found to be heterozygous (16.0 and 17.5 Kb alleles) in three family members, patient 4 and both parents; they have 16.0 and 17.5 Kb alleles as shown in Table I. Patient 5 showed only the 16.0 Kb band and was, therefore, uninformative. In addition, three other loci in 15(q11.2) (D15S18, D15S13 and D15S10) were analyzed for deletion and were also found to be heterozygous, and therefore not deleted in patients 4 and 6. Patient 5 was heterozygous at D15S13, and the other two loci were not tested. While no quantitative dosage blots were done, there were no obvious differences in intensities between the two alleles in the mother that would be indicative of a mosaic deletion at any of the loci tested.

Molecular studies were not informative for the probes utilized for patients in family 2.

Table I. Molecular Data in Family 1

Tissue	Probe	Locus	Father	Mother	Pt.4	Pt.5
			Alleles Sizes in Kb			
Blood	pTD189-1	D15S13	3.5	2.1/3.5	2.1/3.5	2.1/3.5
	pTD3-21	D15S10	7.8/8.9	7.8/8.9	7.8/8.9	n.t.
	pIR10-1	D15S12	16.0/17.5	16.0/17.5	16.0/17.5	16.0
	pIR39	D15S18	8.5/9.0	8.5/9.0	8.5/9.0	n.t.
Skin	pTD189-1	D15S13	*	2.1/3.5	2.1/3.5	*
	pIR10-1	D15S12		16.0/17.5	16.0/17.5	

n.t.=not tested; *tissue was not obtained from these individuals

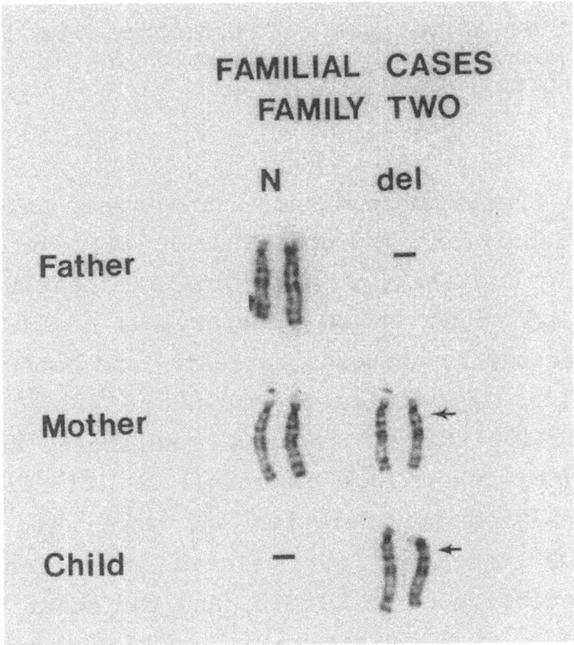

Figure 6. Partial karyotypes of chromosome 15s in members of family 2. Chromosome 15 with the deletion is to the right side (arrow) of its normal homolog.

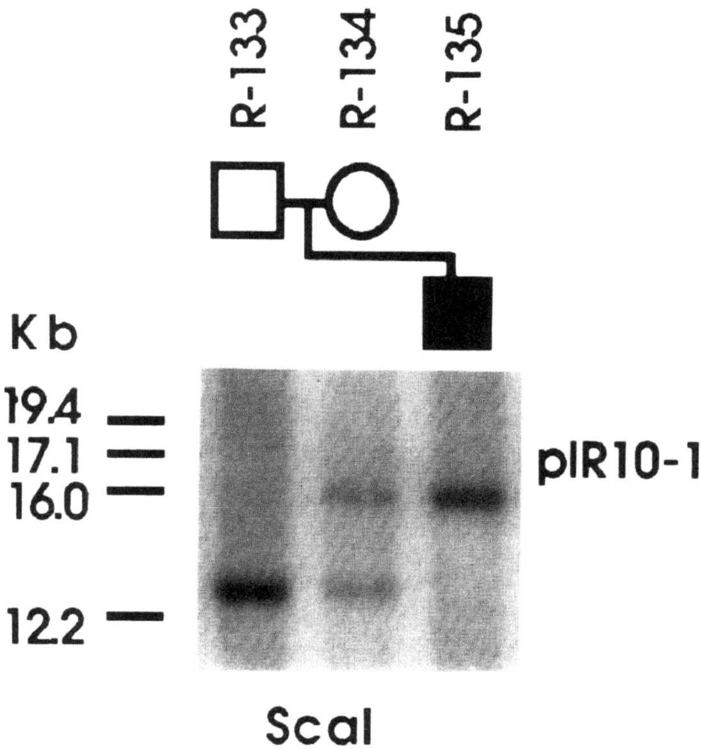

Figure 7. Patient 1 is deleted for D15S12 as revealed by a failure to receive an obligate 12.5 Kb allele from his father. This presumably is not caused by uniparental disomy since the patient has a cytological deletion.

DISCUSSION

Our study includes three patients with mosaicism for 15(q11-13) deletion (patients 1-3) and two families in which the mothers were mosaic for deletion (patients 6 and 8) and their children had deletion in all cells examined (patients 4, 5 and 7). The phenotype of the mothers included a mild degree of mental retardation, with patient 8 also having a borderline personality disorder and mild microcephaly. Patient 6 was of average height with unremarkable features. Her hands were small with hyperextensible finger joints. For comparison with previously reported cases only patients 1-5 and 7 will be considered here.

A deletion of 15(q11-13) has been reported in at least 50% of patients with either PWS or AS. Other patients have been reported with a deletion involving the same region of chromosome 15 but without the clinical phenotype of either PWS or AS. Schwartz et al (1985) and Greenberg and Ledbetter (1987) compiled information on 17 non-classical PWS or AS patients with deletion 15q to determine if a common phenotype could be defined. Schwartz et al (1985) concluded that, in spite of a relatively heterogeneous clinical picture, certain common characteristics could be seen. These included both mental and motor developmental delay, an unusually shaped forehead, malformed ears, ocular anomalies, cryptorchidism, skeletal anomalies, neurological abnormalities (especially severe hypotonia), and Simian creases.

Greenberg and Ledbetter (1987) found that their two patients had physical findings which overlapped those previously described by Schwartz et al (1985), but they also had features not previously noted. Their phenotypic description included broad nasal bridge with telecanthus, full nasal tip, flare of the nasal alae, relatively long philtrum, high-arched palate, and developmental delay. The clinical reports by Reynolds et al (1987) and Galan et al (1991) confirm some common phenotypic findings in patients with non-PWS and non-AS. However, the clinical features of patient 3 in the report by Kaplan et al (1987) were different. She was obese, but her weight gain began at 8 years of age and did not involve inappropriate eating. Her IQ was reported to be normal, but she needed special education classes in school for reading. Her height was at the 25th centile, and her dysmorphic features included triangular mouth and broad neck.

From the clinical description of our patients it is apparent that the phenotype does not fit the description of findings in either PWS or AS. However, there is considerable overlap with the non-PWS and non-AS patients previously described. Comparing the common characteristics of the patients reviewed by Schwartz et al (1985) with those of our patients, developmental delay and large or unusually shaped forehead with broad nasal bridge were features common to all six patients in our study. Five of six patients had hypotonia and one had hypertonia. None of the patients in our study had ocular or skeletal abnormalities. All male patients had normal genitalia. The nasal features described by Greenberg and Ledbetter (1987) were seen in 1/6 of our patients, but the long philtrum was a feature of 5/6 patients. A highly arched palate was noted in 3/6 patients. In addition, 2/6 had mild micrognathia, 3/6 had prominent premaxilla and all patients had delayed or absent speech. None of our patients had developed seizures, whereas both patients described by Greenberg and Ledbetter (1987) and the patient reported by Reynolds et al (1987) had developed seizures before 2 1/2 years.

Head CT on patient 2 in our study revealed ventricular enlargement and equivocal cerebral atrophy. Two other reports (Schwartz et al, 1985; Galan et al, 1991) included the results of cranial ultasound and axial tomography. The patient of Schwartz et al (1985) showed bilateral subdural effusions secondary to cerebral atrophy with a large space between the calvarium and cortex, while the patient of Galan et al (1991) showed cortical brain atrophy with bilateral enlarged ventricles.

The etiology of PWS and AS appears to be complex and is not clearly defined at the present time. As with other contiguous gene syndromes the exact phenotype of PWS and AS will depend largely on the size and the location of a deletion (excluding uniparental disomy). However, unlike most other microdeletion syndromes, the manifestation of PWS or AS is also dependent upon the parental origin of the deleted segment since genomic imprinting is believed to be involved in the expression of the disease (Knoll et al, 1989; Nicholls et al, 1989; Williams et al, 1990; Malcolm et al, 1991). Since DNA segments have not yet been isolated that span the entire 3-4 million base pairs that are estimated to comprise 15(q11.2-q12) it has not been possible to define "critical regions" for these two syndromes, and it is not yet possible to fully discern the sizes and the locations of the breakpoints in our patients. The patients described here who have deletion of one or more loci would be expected to have some clinical overlap with either PWS or AS, depending on the parental origin, if this locus were part of the

PWS or AS critical region. Those patients who have a cytological deletion but no molecular deletion may have deletions that are outside of the region of analysis defined by the DNA probes reported here but might still overlap with a PWS or AS critical region. It will not be possible to determine whether these cases are deviations of PWS and AS or are discrete clinical entities caused by deletion of DNA outside of the PWS or AS critical regions without the development of additional, adjacent DNA markers.

Patient 1 was found to have a paternal deletion of one locus and might therefore be expected to have some clinical overlap with PWS. The familial cases inherited a maternal deletion and might be expected to have some features of AS. The phenotype of their carrier parent could range from normal to a phenotype similar to that of the child's depending of whether their deletion was de novo or inherited, and upon the sex of the contributing parent, if inherited. A report by Niikawa and Hamabe (at this Workshop) described a family in which D15S10 was deleted in three patients with AS and their phenotypically normal mother; the other four loci were intact. The mother inherited this submicroscopic deletion from her father who was also normal. This suggests that a deletion in this region of the genome is not always associated with abnormal phenotype but depends partly on the sex of the contributing parent.

With respect to the existence of mosaicism for deletion in 15(q11Q13), Cassidy et al (1984) found two of their 12 PWS patients to have mosaicism. One patient had deletion in 36% of the cells analyzed and the other in 42%; no deletion was detected in the remaining cells. No major clinical differences were found between patients with mosaicism and those with deletion in all cells. We believe that mosaicism may be a common occurrence and needs to be evaluated further since it may have been overlooked because of technical and interpretative problems.

REFERENCES

Aldridge J, Kunkel L, Burns G, Tantravahi U, Lalande M, Brewster T, Moreau E, Wilson M, Bronley W, Roderick T, Latt SA (1984) A strategy to reveal high frequency RFLP along the human X chromosome in human XX males. Am J Hum Genet 36:546-564

Baraitser M, Patton M, Lam STS, Brett EM, Wilson J (1987) The Angelman (happy puppet) syndrome: Is it autosomal recessive? Clin Genet 31:323-330

Burke CM, Kousseff BG, Gleesen M, O'Connell BM, Devlin JG (1987) Familial Prader-Willi syndrome. Arch Intern Med 147:673-675

Cassidy SB, Thuline HC, Holm VA (1984) Deletion of chromosome 15(q11-13) in a Prader-Labhart-Willi syndrome clinic population. Am J Med Genet 17:485-495

Donlon TA, Lalande M, Wyman A, Bruns G, Latt SA (1986) Isolation of molecular probes associated with the chromosome 15 instability in the Prader-Willi syndrome. Proc Natl Acad Sci (USA) 83:4408-4412

Feinberg AP, Vogelstein B (1983) A technique for radiolabeling DNA restriction endonuclease fragments to high specific activity. Anal Biochem 132:6-13

Galan F, Aguilar MS, Gonzales J, Clemente F, Sanchez R, Tapia M, Moya M (1991) Interstitial deletion without a classic Prader-Willi phenotype. Am J Med Genet 38:532-534

Greenberg F, Ledbetter DH (1987) Deletions of proximal 15q without Prader-Willi syndrome. Am J Med Genet 28:813-820

Kaplan LC, Wharton R, Elias E, Mandell F. Donlon TA, Latt SA (1987) Clinical heterogeneity associated with deletions in the long arm of chromosome 15: Report of 3 new cases and their possible genetic significance. Am J Med Genet 28:813-820

Knoll JHM, Nicholls RD, Magenis RE, Graham JM, Lalande M. Latt SA (1989) Angelman and Prader-Willi syndrome share a common chromosome 15 deletion but differ in parental origin of the deletion. Am J Med Genet 32:285-290

Ledbetter DH, Riccardi VM, Airhart SD, Strobel RJ, Keenan BS, Crawford JD (1981) Deletions of chromosome 15 as a cause of the Prader-Willi syndrome. N Engl J Med 104:325-329

Lubinsky M, Zellweger H, Greenswag LR, Larson G, Hansmann I, Ledbetter DH (1987) Familial Prader-Willi syndrome with apparently normal chromosomes. Am J Med Genet 28:37-43

Magenis RE, Brown MG, Lacey DA, Budden S, LaFranchi S (1987) Is Angelman syndrome an alternate result of del(15)(q11q13)? Am J Med Genet 28:829-838

Malcolm S, Clayton-Smith J, Nichols M, Robb S, Webb T, Armour JAL, Jeffreys AJ, Pembrey ME (1991) Uniparental paternal disomy in Angelman's syndrome. Lancet 337:694-697

Nicholls RD, Knoll JHM, Butler MG, Karam S, Lalande M (1989) Genetic imprinting suggested by maternal heterodisomy in non-deletion Prader-Willi syndrome. Nature 342:281-285

Reynolds JF, Daniel A, Fitzgerald J (1987) Brief clinical report: Atypical phenotype associated with deletion (15)(pter->q11::q13->qter) Am J Med Genet 28:55-58

Schwartz S, Max SR, Panny SR, Cohen MM (1985) Deletions of proximal 15q and nonclassical Prader-Willi syndrome phenotypes Am J Med Genet 20:255-263

Southern E (1975) Detection of specific sequences among DNA fragments separated by gel electrophoresis. J Mol Biol 98:503-517

Williams CA, Zori RT, Stone JW, Gray BA, Cantu ES, Ostrer H (1990) Maternal origin of 15q11-13 deletions in Angelman syndrome suggests a role for genomic imprinting. Am J Med Genet 35:350-353

Willems PJ, Dijkstra I, Brouwer OF, Smit GPA (1987) Recurrence risk in the Angelman ("happy puppet") syndrome. Am J Med Genet 27:773-780

PRADER-WILLI SYNDROME AND ANGELMAN SYNDROME IN TWO FEMALE COUSINS AS A RESULT OF A FAMILIAL TRANSLOCATION

D.F.C.M. Smeets
Department of Human Genetics
University Hospital Nijmegen
P.O. Box 9101
6500 HB, Nijmegen
The Netherlands

M.R. Nelen, B.C.J. Hamel, A.P.T. Smits, H.J.M. Smeets, J.H.M. Bollen and B.A. van Oost

Department of Human Genetics, University Hospital Nijmegen, P.O. Box 9101, 6500 HB, Nijmegen, The Netherlands

INTRODUCTION

Two first cousins, one with Prader-Willi syndrome (PWS) and one with Angelman syndrome (AS), were both found to have familial translocation 6;15. Cytogenetic and molecular analysis indicate that the child with PWS also has a paternal deletion at the Prader-Willi chromosome region, while the child with AS appears to have paternal disomy for this region. The chromosomal translocation involving 15 may have predisposed to the deletion and disomy.

CASE REPORTS, METHODS AND RESULTS:

Recently, a blood sample from a two week old girl was cytogenetically investigated because of extreme hypotonia and severe feeding difficulties. Peripheral lymphocytes from the proband were cultured for 92 hours in medium 199 supplemented with 5% fetal calf serum phytohemagglutinin and antibiotics. After harvesting the cells, chromosome slides were made according to routine procedures and 15 mitoses were analyzed after G-banding with trypsin-Giemsa. In all cells of the proband there were only 45 chromosomes due to a translocation between the most distal part of the short arm of one chromosome 6 and the proximal long arm of one 15. Therefore, at least the short arm and centromere of the translocated chromosome 15 were absent in this girl.

The chromosomes of both her parents and the proband were then investigated in detail. For this we used a specific high resolution G-banding technique with trypsin-basic fuchsin (Scheres et al, 1982). The advantage of this technique, in contrast to most other banding techniques, is the very distinct visualization of band 15q12.

In the mother, a normal 46,XX female karyotype was found. In both the father and his daughter the translocation 6;15 was found, but there were different breakpoints in the translocated chromosome 15. In the father, only the short arm and centromere of the translocated chromosome 15 were absent; in the proband the proximal part of the long arm, including bands 15q11 and q12, was also missing, giving the following chromosome constitutions:

Father: 45,XY,t(6;15)(p25.2;q11),

Proband:45,XX,t(6;15)(p25.3;q13) (Figure 1).

Therefore, the girl inherited a chromosome 15 from her father with a deletion involving the Prader-Willi chromosome region. Such a paternally derived deletion very often leads to PWS which is indeed compatible with the clinical picture of the proband at birth (Ledbetter et al, 1981; Butler and Palmer, 1983; Nicholls et al, 1989; Butler, 1990). This deletion must have arisen in the father as a result of an unequal cross-over during meiosis I, probably induced by incomplete pairing of the homologues within the trivalent figure.

Further family studies revealed several other translocation carriers, among them a mentally retarded girl who was the first cousin of the proband (Figure 2). High resolution chromosome studies showed the presence of exactly the same translocation chromosome in all her cells as was seen in her father and the other normal carriers. In the normal chromosome 15 no deletion of the Prader-Willi chromosome region was found either.

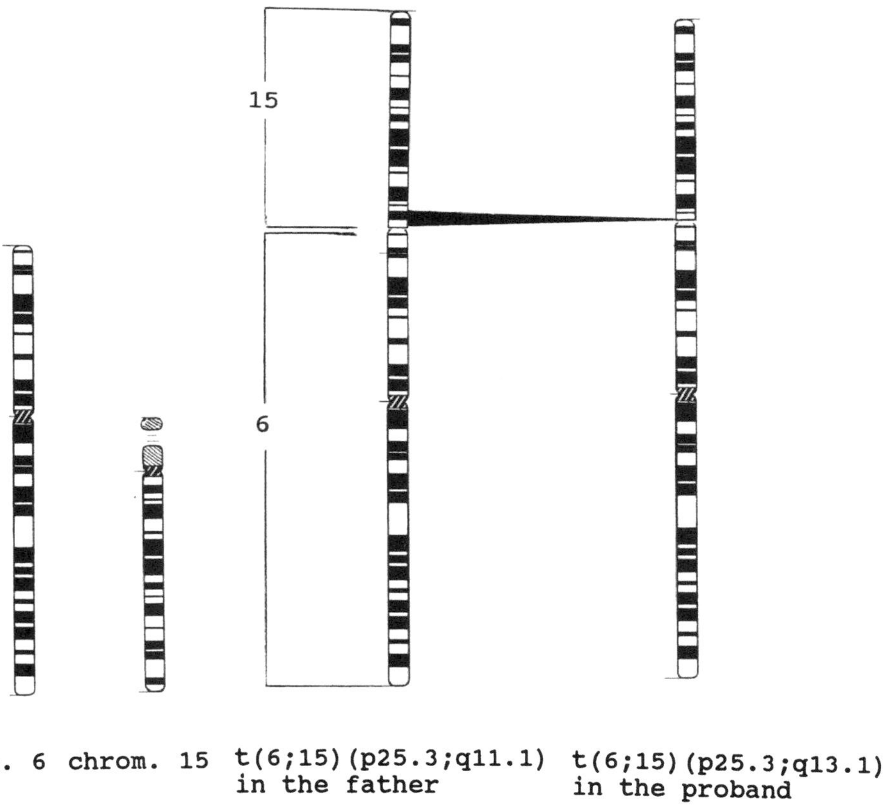

Figure 1: Difference between the translocation chromosome in the proband and her father.

At the same time, DNA studies were performed in this family using a number of polymorphic probes derived from the Prader-Willi/Angelman region (Figure 2). In the proband all probes appeared to fall within the deletion in chromosome 15. Two informative probes lack the paternal signals (a and F), while the remaining four appeared to be deleted based on dosage analysis. Even more interesting were the DNA results in the mentally retarded cousin of the proband, because the maternal alleles of one of the probes (p189.1) were absent. Extensive dosage analysis, however, clearly revealed the presence of two copies (a) of this allele, indicating that two paternal alleles were present.

During these investigations, this girl was examined by two clinical geneticists, both of whom made the same diagnosis:

Angelman syndrome. Uniparental paternal heterodisomy for chromosome 15 is likely in this girl, based upon the DNA results and the fact that her father carries the translocation 6;15, which sharply increases the risk for nondisjunction during meiosis I (adjacent instead of alternate) leading to sperm which are disomic and nullisomic for chromosome 15. All other chromosome 15 markers were compatible with this interpretation. In this case, a sperm cell disomic for chromosome 15 might have fertilized an egg cell lacking a chromosome 15, or the maternal chromosome 15 may have vanished during one of the first cell divisions.

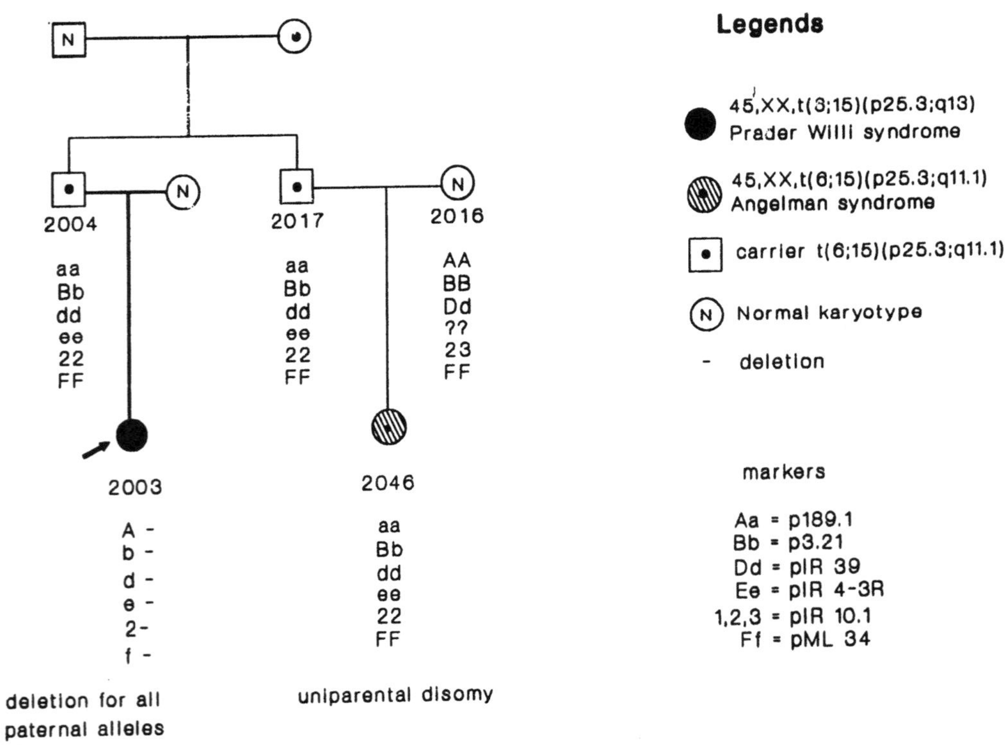

Figure 2: Part of the family pedigree including the cytogenetic and molecular data.

Although uniparental disomy is not frequent in Angelman syndrome (Knoll et al, 1991), it has been reported in a few cases

(Malcolm et al. 1991). However, a structural chromosome aberration involving chromosome 15 may dramatically increase the risk for disomy due, for example, to nondisjunction (Schinzel, 1991).

CONCLUSION:

In a small family in which a translocation 6;15 is segregating, we identified a patient with Prader-Willi syndrome with a paternally derived deletion in chromosome 15 and a first cousin of this proband with Angelman syndrome most likely the consequence of uniparental paternal heterodisomy for chromosome 15 In our opinion, the occurrence of both syndromes in this family can be directly related to the presence of the 6;15 translocation, indicating that the existence of such a structural chromosome aberration may greatly increase the risk of having affected offspring.

REFERENCES

Butler MG (1990) Prader-Willi syndrome: Current understanding of cause and diagnosis. Am J Med Genet 35:319-332

Butler MG, Palmer CG (1983) Parental origin of chromosome 15 deletions in Prader-Willi syndrome. Lancet 1:1285-1286

Knoll JHM, Glatt KA, Nicholls RD, Malcolm S, Lalande M (1991) Chromosome 15 uniparental disomy is not frequent in Angelman syndrome. Am J Hum Genet 48:16-21

Ledbetter DH, Riccardi VM, Airhart SD, Strobel RJ, Keenan BS, Crawford JD (1981) Deletions of chromosome 15 as a cause of the Prader-Willi syndrome. N Engl J Med 304:325-329

Malcolm S, Clayton-Smith J, Michols M, Robb S, Webb T, Armour JAL, Jeffreys AJ, Pembrey ME (1991) Uniparental paternal disomy in Angelman's syndrome. Lancet 337:694-697

Nicholls RD, Knoll JH, Glatt K, Hersh JH, Brewster TD, Graham JM Jr, Wurster-Hill D, Wharton R, Latt SA (1989) Restriction fragment length polymorphisms within proximal 15q and their use in molecular cytogenetics and the Prader-Willi syndrome. Am J Med Genet 33:66-77

Scheres JMJC, Merkx GFM, Hustinc TWJ (1982) Prometaphase banding of human chromosomes with basic fuchsin. Hum Genet 61:8-11

Schinzel AA (1991) Uniparental disomy and gene localization. Am J Hum Genet 48:424-425

IMPLICATIONS FOR THE RECURRENCE RISK IN THE PRADER-WILLI SYNDROME ON THE BASIS OF PROPOSED GENOMIC IMPRINTING

Ingo Kennerknecht
Abteilung Klinische Genetik de Universität
Frauenstrasse 29
D-7900 Ulm
Germany

INTRODUCTION

The Prader-Willi syndrome (PWS) describes a complex malformation syndrome with abnormal behavior and intellectual deficiency. About 60% of the patients show a deletion in the PW-chromosome-region 15q11-13 (PWCR) (review Ledbetter et al, 1987). In less than 5% of patients other chromosome aberrations are found, and in nearly 40% there are no cytogenetically detectable deletions. In contrast, there are deletions in PWCR reported with a normal phenotype and in Angelman syndrome (AS) (Ledbetter et al, 1987). Murdock and Wurster-Hill (1986) list a number of proposed pathogenetic mechanisms. As the de novo-deletion-chromosome 15 derives paternally in PWS, in contrast to maternal origin in AS, a genomic imprinting mechanism was postulated. This was further supported by the surprising finding of maternal heterodisomy for PWCR in patients not showing a deletion (Nicholls et al, 1989). PWS usually occurs sporadically, yet familial recurrence is occasionally reported. It is shown that the sporadic and familial cases are consistent with a unique genetic model involving genomic imprinting.

MATERIALS AND METHODS

Published data of approximately 800 patients with PWS, including 7 families with recurrence of PWS, were examined. Only those pedigrees in which the patients were consistent with the diagnostic criteria collected and evaluated by Cassidy (1987b)

NATO ASI Series, Vol. H 61
Prader-Willi Syndrome
Edited by S. B. Cassidy

were selected. Weinberg's sib- and proband method was used for a simple correction of a possible ascertainment bias.

RESULTS AND DISCUSSION

Sporadic cases are due to the absence of paternal PWCR either by deletion or by maternal heterodisomy (Nicholls et al, 1989). Alternately, PWS might be caused not only by absence, but also by dysfunction (e.g., by "PWS" gene mutation?) of the paternal PWCR. This might be true for familial cases. There are no reports of familial cases with an interstitial deletion.

The following proposed mechanism for familial manifestation of PWS is based on the hypothesis of maternal imprinting and paternal erasure of the imprint. It is further assumed that imprinting of the "dysfunctional" PWS gene/region during female meiosis prevents PWS, whereas erasure of imprinting in male transmitters would result in 50% affected children. To date no PWS patient has reproduced. Thus, transmission happens only through normal carriers, who received the dysfunctional but imprinted PWCR from their mother. As PWCR is imprinted in these carriers through their maternal meiosis, PWS is not manifest. However, their gametogenesis is accompanied by an erasure of the imprint. If the chromosome with the dysfunctional PWCR is transmitted (50% chance), the child will express PWS. This is compatible with the unexpectedly high segregation ratio of about 0.5 derived form published pedigrees. This model would also allow concurrence of affected cousins, as was observed by Clarren and Smith (1977).

For genetic counseling this mechanism would suggest different recurrence risk estimations depending on the number of affected sibs. If only one child is affected by PWS it is not possible to distinguish between a sporadic or an isolated familial case. For this situation I evaluated from literature data a recurrence risk of 0.4%. This was derived as follows. Three published pedigrees show recurrence of PWS (Burke et al,

1987; Ishikawa et al, 1987; Lubinsky et al, 1987). In these three pedigrees there are a total of 16 sibs, 9 of whom have manifestations of PWS. Excluding one proband per pedigree, 6 sibs with PWS result. For the calculation of the overall recurrence risk, the reported number of 800 isolated sporadic cases was used. Since full sibship data is not available for these families, the number of siblings was extrapolated based on data collected by Clarren and Smith (1977) from 72 families with a single PWS proband and a total of 132 children. This results in an average of 1.8 children per family, or an estimated 1440 siblings for the 800 reported families with a single PWS proband. Hence, the estimated frequency of affected sibs is 6/1440, or 0.42%. Estimations by others range from <0.1% (Cassidy, 1987a) to 1.6% (Clarren and Smith, 1977). In cases of two (or more) affected sibs a 50% recurrence risk might be postulated.

The mode of genomic imprinting suggested here is still hypothetical until proven or disproven by molecular genetic findings.

REFERENCES

Burke CM, Kouseff BG, Gleeson M, O'Connell BM, Devlin JG (1987) Familial Prader-Willi syndrome. Arch Intern Med 147:673-675

Cassidy SB (1987a) Recurrence risk in Prader-Willi syndrome (letter). Am J Med Genet 28:59-60

Cassidy SB (1987b) Prader-Willi syndrome: Characteristics management, and etiology. Ala J Med Sci 24:169-175

Clarren SK, Smith DW (1977) Prader-Willi syndrome: Variable severity and recurrence risk. Am J Dis Child 131:789-800

Ishikawa T, Kanayama M, Wada Y (1987) Prader-Willi syndrome in two siblings: One with normal karyotype, one with a terminal deletion of distal Xq. Clin Genet 32:295-299

Ledbetter DH, Greenberg F, Holm VA, Cassidy SB (1987) Conference report: Second annual Prader-Willi syndrome scientific conference. Am J Med Genet 28:779-790

Lubinsky W, Zellweger H, Greenswag L, Larson G, Hansmann I, Ledbetter D (1987) Familial Prader-Willi syndrome with apparently normal chromosomes. Am J Med Genet 28:37-43

Murdock RL, Wurster-Hill DH (1986) Non-reciprocal translocation (5;15), isodicentric (15) and Prader-Willi syndrome. Am J Med Genet 25:61-69

Nicholls RD, Knoll JH, Glatt K, Hersh JH, Brewster RD, Graham JM Jr, Wurster-Hill D, Wharton R, Latt SA (1989) Restriction fragment length polymorphisms within proximal 15q and their use in molecular cytogenetics and the Prader-Willi syndrome. Am J Med Genet 33:66-77

DIAGNOSTIC CRITERIA FOR PRADER-WILLI SYNDROME

Vanja A. Holm, M.D.
Department of Pediatrics
University of Washington School of Medicine
Child Development and Mental Retardation Center WJ-10
University of Washington
Seattle, Washington 98195
USA

Suzanne B. Cassidy, M.D.[1], Merlin G. Butler, M.D., Ph.D.[2], Jeanne M. Hanchett, M.D.[3], Frank Greenberg[4], Barbara Y. Whitman[5], and Louise R. Greenswag, R.N., Ph.D.[6].

University of Arizona College of Medicine, Tucson, Arizona[1]; Vanderbilt University, Nashville, Tennessee[2]; The Rehabilitation Institute of Pittsburgh, Pittsburgh, Pennsylvania[3]; Baylor College of Medicine, Houston, Texas[4]; Cardinal Glennon Hospital, St. Louis, Missouri[5]; The University of Iowa, Iowa City, Iowa[6]

INTRODUCTION

Most experienced clinicians agree that the diagnosis of Prader-Willi syndrome (PWS) remains a clinical one. The 15q deletion has been considered confirmatory when present. However, several studies have established that only 50 - 70% of the patients with the syndrome show this finding, always in the paternally derived chromosome 15. Some patients with the PWS have shown other abnormalities of chromosome 15 (Cassidy and Ledbetter, 1989). In addition, it is now well established that Angelman syndrome is also associated with the same deletion of a chromosome 15 inherited from the mother rather than the father (Magenis et al, 1990). Finally, patients with other conditions, e.g., Williams syndrome, and hypotonia of unknown etiology have also been reported to demonstrate the 15q deletion on cytogenetic studies (Kaplan et al, 1987; Holm, 1991).

Thus, PWS is not a laboratory but a clinical diagnosis. At this time, clinicians use their experience when deciding whether or not an individual patient fits into this diagnostic group.

NATO ASI Series, Vol. H 61
Prader-Willi Syndrome
Edited by S. B. Cassidy

This can lead to diagnostic inaccuracy, which is not only confusing for families, but can result in erroneous research results due to inclusion of misdiagnosed patients. Recently a laudable trend has appeared towards developing generally acceptable diagnostic criteria for genetic syndromes without undisputable diagnostic biological markers through consensus of informed physicians. This has been accomplished for neurofibromatosis, Rett syndrome, and tuberous sclerosis (National Institutes of Health, 1988; the Rett Syndrome Diagnostic Criteria Work Group, 1988; Gomez, 1988). We have held a consensus workshop to develop diagnostic criteria for PWS, and have tested it on over 100 patients referred to the workshop physicians for a possible diagnosis of PWS.

METHODS

Five physicians, a nurse, and a social worker active during the past five years in the Prader-Willi Syndrome Association (USA) annual meetings and the associated scientific meeting met in January, 1991, and developed consensus diagnostic criteria for the diagnosis of the syndrome. Methods were applied which had been used for other genetic disorders, and previously proposed diagnostic criteria for the syndrome (Hall and Smith, 1972; Holm, 1981) were considered. The developed guidelines were subsequently tested on 113 patients by clinicians involved in the consensus workshop.

The symptoms of PWS change dramatically with age. Therefore, the criteria developed for diagnosis were divided into two different age groups: 3 years to adulthood, and 0 to 36 months. Three categories of criteria evolved: major, minor, and supportive. A scoring system was developed whereby major criteria were scored as one point (with one exception as indicated below), and minor criteria for 1/2 point. Supportive criteria were not included in the point system, but were intended to give greater confidence to the diagnostician.

PATIENT POPULATION

In the testing process, the participating clinicians were asked to score their referred or previously evaluated patients and judge whether, in their experience, a patient had classical PWS, atypical PWS, did not have PWS, or whether they were unsure. Patients' sex and age were indicated on the scoring sheet, which also included information as to whether patients were examined at the time of scoring, or if the information was obtained through review of charts of patients previously examined.

There were 113 patients from five different centers tested using the developed criteria. Information on 96, or 85%, was obtained through chart review of patients previously evaluated by the authors; 17, or 15%, were examined at the time of scoring. Forty-eight percent were males, 52 percent females. Age of the 26 children in the younger group ranged from 0.5 to 3 years with a mean age of 1.9 years. In the older group there were 87 patients, and age ranged from 3.3 to 38.6 years with a mean of 14.5 years. The mean of the total group was 11.6 years.

RESULTS

In the 3 year to adulthood group, the consensus process led to the development of 8 major, 10 minor and 9 supportive criteria (Table I). Eight points are deemed necessary for diagnosis, and 5 of those must come from the major criteria group.

In the 0 to 36 month group, consensus led to 7 major, no minor and 6 supportive criteria (Table II). Items 1-6 were given a score of 1, with a score of 2 for item 7. To qualify for a diagnosis a child must have 6 points.

Table I. Initial Proposed Diagnostic Criteria, Age Three Years to Adulthood.

Major Criteria

1. History of neonatal central hypotonia and poor suck with later improvement.
2. Obesity (primarily central) beginning between ages 1 and 6 without intervention.
3. Mild to moderate mental retardation or learning problems, or global developmental delay in a child less than 6 years of age.
4. Characteristic behavior problems - temper tantrums, violent outbursts, obsessions and compulsiveness, argumentativeness, perseverativeness, oppositional behavior, rigidity, manipulativeness, possessiveness, stealing, lying, stubbornness.
5. Short stature for genetic background by age 15 (without intervention).
6. Characteristic facial features - narrow bifrontal diameter, thin upper lip, down-turned mouth, almond-shaped eyes.
7. Hypogonadism - with any of the following:
   a. Genital hypoplasia:
      Male: scrotal hypoplasia, cryptorchidism, small penis and testes.
      Female: labia minora and/or clitoral hypoplasia
   b. Incomplete pubertal development and scant pubic hair without intervention after 16 years of age.
8. Deletion 15q11-13 on high resolution (>550 bands) or other cytogenetic/molecular abnormality of the Prader-Willi chromosome region (PWCR), including maternal disomy.

Minor Criteria

1. History of decreased fetal movement or infantile lethargy or weak cry, improving with age.
2. History of feeding problems and poor weight gain in infancy.
3. Hyperphagia/food foraging/obsession with food.
4. Decreased vomiting.
5. Sleep disturbance or sleep apnea.
6. Skin picking.
7. Articulation defects.
8. Small hands and feet (hands <25%, feet <10% for height).
9. Thick viscous saliva with crusting at corners of the mouth.
10. Normal neuromuscular studies.

Supportive Findings

1. Hypopigmentation - fair skin and hair compared to family.
2. Esotropia and/or myopia.
3. High pain threshold.
4. Altered temperature sensitivity.
5. Scoliosis and/or kyphosis.
6. Early adrenarche.
7. Osteoporosis
8. Narrow hands with straight ulnar border.
9. Unusual skill with jigsaw puzzles.

Table II. Initial Proposed Diagnostic Criteria, Age 0 to 36 Months.

Major Criteria

1. Neonatal and infantile central hypotonia with poor suck and lethargy, gradually improving with age.
2. Feeding problems with need for special feeding techniques and poor weight gain/failure to thrive.
3. Characteristic facies with dolichocephaly, almond-shaped eyes, narrow face, small-appearing mouth, down-turned corners of mouth.
4. Genital hypoplasia:
   Female: hypoplasia of labia minora and/or clitoris.
   Male: cryptorchidism, small penis, scrotal hypoplasia.
5. Global developmental delay.
6. Excessive or rapid weight gain on weight-for-length chart after 12 months (excessive is defined as crossing two centile channels).
7. Deletion 15q11-13 on high resolution (>550 bands) or other cytogenetic/molecular abnormality of the Prader-Willi chromosome region (PWCR), including maternal disomy.

Supportive Findings

1. Decreased fetal movement.
2. Weak cry <6 months.
3. Lethargy <6 months.
4. Temperature instability.
5. Esotropia.
6. Negative neuromuscular studies.

Clinical diagnosis and point scores for tested patients in the two age groups are displayed in Table III.

Table III. Clinical Diagnosis and Initial Proposed Criteria Scores for Tested Patients.

	No	Total Score		Major Criteria	
		Range	Mean	Range	Mean
3 Years to Adult					
Typical PWS	65	8-12.5	10.1	5-8	6.9
Not PWS	11	1.5-6	4.3	1-5	3.4
Atypical PWS or Unsure	11	3-10	7.0	3-6	4.9
0 to 36 Months					
Typical PWS	15	5-8	6.7	same	
Not PWS	7	0-5	2.6	same	
Atypical PWS or Unsure	4	4-5	4.3	same	

The percentage of people in the two age groups who showed the major and minor criteria and supportive findings is displayed by diagnostic categories in Tables IV and V.

Table IV. Major and Minor Criteria and Supportive Findings, Three Years to Adulthood.

	Typical PWS N=65 (%)	Not PWS N=11 (%)	Atypical PWS N=11 (%)
Major Criteria			
1. Hypotonia	96.9	27.3	90.9
2. Obesity	96.9	81.8	90.9
3. MR,LD,DD	93.8	90.9	81.8
4. Behavior Problems	84.6	63.6	72.7
5. Short Stature	75.4	45.5	45.5
6. Dysmorphism	100.0	9.1	45.5
7. Hypogonadism	87.7	9.1	72.7
8. PWCR Abnormality	75.4	9.1	18.2
Minor Criteria			
1. Poor Fetal Activity	84.6	27.3	45.5
2. Feeding Problems	96.9	18.2	72.7
3. Hyperphagia	87.7	45.5	54.5
4. Decreased Vomiting	9.2	0.0	18.2
5. Sleep Disturbance	43.1	0.0	9.1
6. Skin Picking	75.4	18.2	61.5
7. Articulation Problem	76.9	36.4	61.5
8. Small Hands and Feet	70.8	0.0	61.5
9. Viscous Saliva	66.2	0.0	18.2
10. Normal Muscle Studies	13.8	9.1	18.2
Supportive Findings			
1. Hypopigmentation	41.5	9.1	18.2
2. Eye Abnormality	66.2	9.1	27.3
3. High Pain Threshold	9.2	9.1	9.1
4. Abnormal Temperature Sensitivity	4.6	0.0	0.0
5. Scoliosis/Kyphosis	24.6	9.1	36.4
6. Premature Adrenarche	1.5	0.0	18.2
7. Osteoporosis	3.1	0.0	0.0
8. Straight Ulnar Border	69.2	9.1	27.3
9. Jigsaw Puzzle Skill	6.2	0.0	0.0

MR=Mental retardation; LD=Learning disability; DD=Developmental delay

Table V. Major Criteria and Supportive Findings, Age 0 to 36 Months.

	Typical PWS N=15 (%)	Not PWS N=7 (%)	Atypical PWS N=4 (%)
Major Criteria			
1. Hypotonia	100.0	42.9	75.0
2. Feeding Problems	100.0	14.3	50.0
3. Dysmorphism	100.0	28.6	50.0
4. Genital Hypoplasia	86.5	14.3	100.0
5. Developmental Delay	100.0	28.6	100.0
6. Excess Weight Gain	53.3	14.3	50.0
7. PWCR Abnormality	73.3	57.1	100.0
Supportive Findings			
1. Poor Fetal Activity	42.9	0.0	0.0
2. Weak Cry	64.3	14.3	50.0
3. Lethargy	14.3	0.0	50.0
4. Temperature Instability	0.0	0.0	0.0
5. Esotropia	42.9	14.4	50.0
6. Negative Muscle Studies	57.9	14.3	75.0

DISCUSSION

As shown in Table III, there was good agreement between diagnosis based on clinical experience and the consensus scoring system in the age group 3 years to adulthood. The total score was 8 or above in all the tested typical PWS patients, with 5 or more of those being from the major criteria. In the non-PWS group, the total score was 6 or less in all cases.

The diagnostic criteria for the 0 to 36 months age group had been designed so that all the other 6 criteria had to be present to reach a diagnosis if there was no PWCR abnormality known. As a result, it would not be possible to diagnose PWS in a young infant before onset of obesity if a chromosomal or molecular genetic abnormality was not known to be present. As a consequence, the lower end of the range for total score was 5, less than the stipulated 6, in the typical young PWS group (Table III). There was one child without a PWCR abnormality who was clinically thought to have the syndrome but who only reached 5 points: he was 7 months old and had not begun to gain weight rapidly, confirming that this system will not confirm the diagnosis in very young infants in the absence of chromosome studies or with normal chromosome studies in the absence of molecular studies.

Not surprisingly, obesity, mental retardation/learning disability, and behavior problems did not separate typical PWS from non-PWS (Table IV) even though these symptoms are considered essential for diagnosis in older individuals. Patients referred for a diagnosis of PWS tend to be those who are obese and retarded with behavior problems. On the other hand, the characteristic dysmorphic facial features, hypogonadism and early hypotonia had high prevalence in the patients with clinically typical PWS but occurred with low frequency in the non-PWS group.

Hyperphagia, which is a minor criterion, was somewhat more common than behavior problems, a major criterion, and less prevalent among the patients felt not to have the syndrome (Table IV). Short stature had the lowest prevalence among the major criteria and was relatively common among the non-PWS patients. On the other hand, history of early feeding problems (a minor criterion) had a high prevalence in typical PWS but a relatively low prevalence in the non-PWS group. Decreased fetal activity was likewise a very frequent minor criterion. As this is such a subjective finding, and is most likely secondary to hypotonia, it was difficult to ascertain the importance of this finding to diagnosis.

Two minor criteria in the 3 year to adulthood group with very low prevalence were decreased vomiting and normal neuromuscular studies. On the other hand, three supportive findings, narrow hands with straight ulnar border, strabismus/myopia, and hypopigmentation, had high prevalence in the typical PWS and low prevalence in the non-PWS group.

The seven major criteria in the 0-36 months age group were quite prevalent among patients with typical PWS with the exception of excessive weight gain (53.3%). Instead, four supportive criteria were found to be prevalent: weak cry, neuromuscular studies, esotropia and decreased fetal activity. The last finding is difficult to evaluate, as previously discussed. It was also noted that two of the testing clinicians had added scoliosis as an additional supportive finding in their young patients.

The percentage of criteria exhibited by patients with atypical PWS and the patients with unsure diagnosis fell between the typical and non-PWS groups. In this study, there were too few patients to ascertain the significance of the numbers obtained.

As a consequence of the testing process, therefore, the following changes were made in the proposed criteria system for the diagnosis of PWS:

3 year to adulthood

1. Hyperphagia was changed from a minor to a major criterion.
2. Behavior problems was changed from a major to a minor criterion.
3. Early feeding problems was changed from a minor to a major criterion.
4. Short stature was changed from a major to a minor criterion.
5. Decreased vomiting and normal neuromuscular studies were changed from minor criteria to supportive findings.
6. Narrow hands with straight ulnar border, eye abnormalities (strabismus/myopia), and hypopigmentation were changed from supportive findings to minor criteria.

0 to 36 months age group

7. A new Minor Criteria category was added for this age group, with 1/2 point each: weak cry, esotropia, and negative neuromuscular studies.
8. Scoliosis was added to the supportive findings.

The first six recommendations have been adjusted to recognize the prevalence information uncovered in this study of 113 patients. Recommendation 7 is based on careful clinical observations made by the experienced physicians participating in this study. This change will make it possible to diagnose PWS in infants before excessive weight gain occurs. The last recommendation acknowledges the observation made by some of the participating physicians.

The final recommendation of the consensus workshop for diagnostic criteria for PWS are listed in Tables VI and VII.

These revised Diagnostic Criteria for PWS represent a vital step toward an objective means of diagnosis or exclusion of diagnosis while clinicians await more definitive and complete laboratory testing. It is expected that they will be both an aid to clinicians and an assistance to researchers seeking improved purity of patient population for inclusion in clinical and laboratory studies.

Table VI. Revised Diagnostic Criteria for PWS: Ages 3 Years and Above.

SCORING: Major Criteria: Valued at 1 point; Minor Criteria: Valued at 1/2 point. A total of 8 points, of which 5 come from major criteria, are required for confident diagnosis of PWS. Supportive Findings have no scoring value.

MAJOR CRITERIA:

1. History of neonatal central hypotonia and poor suck with later improvement
2. History of feeding problems and poor weight gain in infancy
3. Obesity (primarily central) beginning between ages 1 and 6 years without intervention
4. Characteristic facial features: narrow bifrontal diameter, almond-shaped palpebral fissures, thin upper lip, down-turned mouth
5. Hypogonadism, with any of the following:
   (a) Genital hypoplasia:
   Males: Cryptorchidism, small penis and testes for age (<5th centile), scrotal hypoplasia
   Females: Hypoplasia of labia minora and/or clitoris
   (b) Incomplete pubertal development without intervention after age 16 years:

Males: Decreased facial and body hair, lack of voice change
Females: Amenorrhea/oligomenorrhea

6. Mild to moderate mental retardation or multiple learning problems, or global developmental delay in a child <6 years
7. Hyperphagia/food foraging/obsession with food
8. Deletion 15q11-13 on high resolution karyotype (>550 bands) or other cytogenetic/molecular abnormality of PWCR, including maternal disomy.

MINOR CRITERIA:

1. History of decreased fetal movement or infantile lethargy or weak cry, improving with age
2. Characteristic behavior problems: >5 of the following: temper tantrums, violent outbursts, obsessive/compulsive behavior; tendancy to be argumentative, oppositional, rigid, manipulative, possessive, stubborn; perseverates, steals, lies
3. Sleep disturbance or sleep apnea
4. Short stature for genetic background by age 15 without intervention
5. Hypopigmentation: fair skin and hair compared to family
6. Small hands (<25th centile) and/or feet (<10th centile) for height age
7. Narrow hands with straight ulnar border
8. Eye abnormalities (esotropia, myopia)
9. Thick viscous saliva with crusting at corners of the mouth
10. Articulation defect
11. Skin picking

SUPPORTIVE FINDINGS:

1. High pain threshold
2. Decreased vomiting
3. Altered temperature sensitivity
4. Scoliosis and/or kyphosis
5. Early adrenarche
6. Osteoporosis
7. Unusual skill with jigsaw puzzles
8. Normal neuromuscular studies

Table VII. Revised Diagnostic Criteria for PWS: Ages 0 - 36 Months.

SCORING: Major Criteria: Items 1 - 6 are valued at 1 point each, item 7 at 2 points. Minor Criteria: Valued at 1/2 point each. A total of 6 points are required for confident diagnosis of PWS. Supportive Findings have no scoring value.

MAJOR CRITERIA:

1. Neonatal and infantile central hypotonia with poor suck and lethargy, improving with age
2. Feeding problems with need for special feeding techniques and poor weight gain/failure to thrive
3. Characteristic facies with dolichocephaly, almond-shaped eyes, narrow face, small appearing mouth with down-turned corners
4. Genital hypoplasia:
   females: hypoplasia of labia minora and/or clitoris
   males: cryptorchidism, small penis, scrotal hypoplasia
5. Global developmental delay
6. Excessive or rapid weight gain on weight-for-length chart after 12 months (crossing two or more centile channels)
7. Deletion 15q11-13 on high resolution karyotype (>550 bands) or other cytogenetic/molecular abnormality of the PWCR, including maternal disomy

MINOR CRITERIA:

1. Weak cry below 6 months
2. Esotropia
3. Negative neuromuscular studies

SUPPORTIVE FINDINGS:

1. Decreased fetal movement
2. Lethargy below 6 months
3. Temperature instability
4. Scoliosis

REFERENCES

Cassidy SB, Ledbetter DH (1989) Prader-Willi syndrome. Neurol Clin 7:37-54

Gomez MR (1988) Tuberous Sclerosis, 2nd ed. Raven Press, New York

Hall BD, Smith DW (1972) Prader-Willi syndrome: A resume of 32 cases including an instance of affected first cousins, one of whom is of normal stature and intelligence. J Pediat 81:286-293

Holm VA (1981) The diagnosis of Prader-Willi syndrome. In: Holm VA, Sulzbacher S, Pipes PL (eds) The Prader-Willi Syndrome. University Park Press, Baltimore, p 27-36

Holm VA (1991) 15q deletion without the Prader-Willi syndrome developmental phenotype in five young children. (Abstract) Am J Med Genet (in press)

Kaplan LC, Wharton R, Elias E, Mandell F, Donlon T, Latt SA (1987) Clinical heterogeneity associated with deletions of the long arm of chromosome 15: Report of three new cases and their possible genetic significance. Am J Med Genet 28:45-53

Magenis RE, Toth-Fejel S, Allen LJ, Black M, Brown MG, Budden S, Cohen R, Friedman JM, Kalousek D, Zonana J, Lacy D, LaFranchi S, Lahr M, MacFarlane J, Williams CPS (1990) Comparison of the 15q deletions in Prader-Willi and Angelman syndromes: Specific regions, extent of deletions, parental origin, and clinical consequences. Am J Med Genet 35:333-349

National Institutes of Health Consensus Development Conference (1988) Neurofibromatosis. Arch Neurol 45:575-578

The Rett Syndrome Diagnostic Criteria Work Group (1988) Diagnostic criteria for Rett syndrome. Ann Neurol 23:425-428

ACKNOWLEDGEMENTS

The Consensus Workshop members are grateful to PWSA for its financial and emotional support of this effort.

AN AUSTRALIAN COLLABORATIVE STUDY OF PRADER-WILLI SYNDROME INDIVIDUALS AND THEIR FAMILIES

Arabella Smith
Cytogenetics Laboratory
The Children's Hospital
Camperdown
Sydney, New South Wales
Australia

S. White[2], G. Warne[3], P. Montgomery[4], J. Nelson[5], H. Beange[6], J. Pearn[7], J. Henderson[8], R.J. Trent[9]

Cytogenetics Unit, Oliver Latham Laboratory, Department of Health, New South Wales and Children's Hospital, Sydney, Australia[1]; Department of Medical Genetics, Adelaid Children's Hospital, Australia[2]; Endocrinology and Diabetes, Children's Hospital, Melbourne, Australia[3]; Irrabeena, Perth, Australia[4]; Department of Pediatrics, Westmead Hospital, Sydney, Australia[5]; Royal North Shore Hospital, Sydney, Australia[6]; Department of Child Health, University of Queensland, Australia[7]; Douglas Parker Rehabilitation Centre, Hobart, Tasmania, Australia[8]; Department of Molecular Genetics, Royal Prince Alfred Hospital and the University of Sydney, Australia[9]

INTRODUCTION

Prader-Willi syndrome (PWS) is considered to be a rare disorder. Its prevalence in Australia is not known, but the figure of 1 in 10,000 (Holm 1981) is sometimes used. While it is a sporadic disorder, recurrence risks ranging from 0.1% (Cassidy, 1987) to 1-3% (Clarren and Smith, 1977) have been quoted. Family data, including information on siblings, has been published for the American and Canadian populations (Dunn, 1968; Greenswag, 1987). This is a survey of families ascertained because of an individual with PWS - henceforth referred to as PWSF - in Australia.

METHODS

Cases were ascertained through the PWS Association of Australia, PWS clinics, genetics clinics, Developmental Disability

NATO ASI Series, Vol. H 61
Prader-Willi Syndrome
Edited by S. B. Cassidy

Services, pediatricians, endocrinologists and other referrals in each state of Australia. Most of the patients have been previously known (Smith and Simpson, 1982). Results were compiled on the basis of interviews, medical records and questionnaires.

The questionnaire was simple and asked for date of birth of the individual with PWS, the parents and the siblings, sex of siblings, consanguinity, pregnancy loss (either spontaneous abortion [SA], stillbirth [SB] or neonatal death [NND]) and recurrence within the family.

The diagnosis of PWS followed the criteria of Holm (1981). All 6 of the following features were required for inclusion in the study: neonatal hypotonia, hypogonadism, short stature, obesity, characteristic facial appearance, and intellectual disability. Twelve cases have been excluded, 6 because they lacked the facial appearance, 3 because the early history was not typical and 3 because their height was between 25th and 50th percentile. No case under 3 years was included.

Cytogenetic studies were performed in different laboratories in the major city of each state. Many karyotypes of the older patients have been reviewed by one of us (AS). The criteria for "high resolution" varied somewhat between laboratories. Some of the parents of children with interstitial deletion 15q were tested, but this was not uniformly done and data on parental chromosomes are not included in this survey.

In the analysis of the data presented here, cases have not been divided into those with an interstitial deletion and those without, despite much discussion on cytogenetic heterogeneity and its implications (Smith et al, 1990; Ledbetter et al, 1987).

RESULTS

Patients Ascertained: The data presented derive from 77 families who returned questionnaires, 19 who were already known plus 48 ascertained by interviews/medical records. This survey thus covers 144 PWSF, 74 with a male and 70 with a female proband. The ages ranged from 3.5 to 41 years (Figure 1). Overall 110/144 (76.4%) were born before 1980. In 6 families the child with PWS was adopted or in foster care and not all the information sought was available.

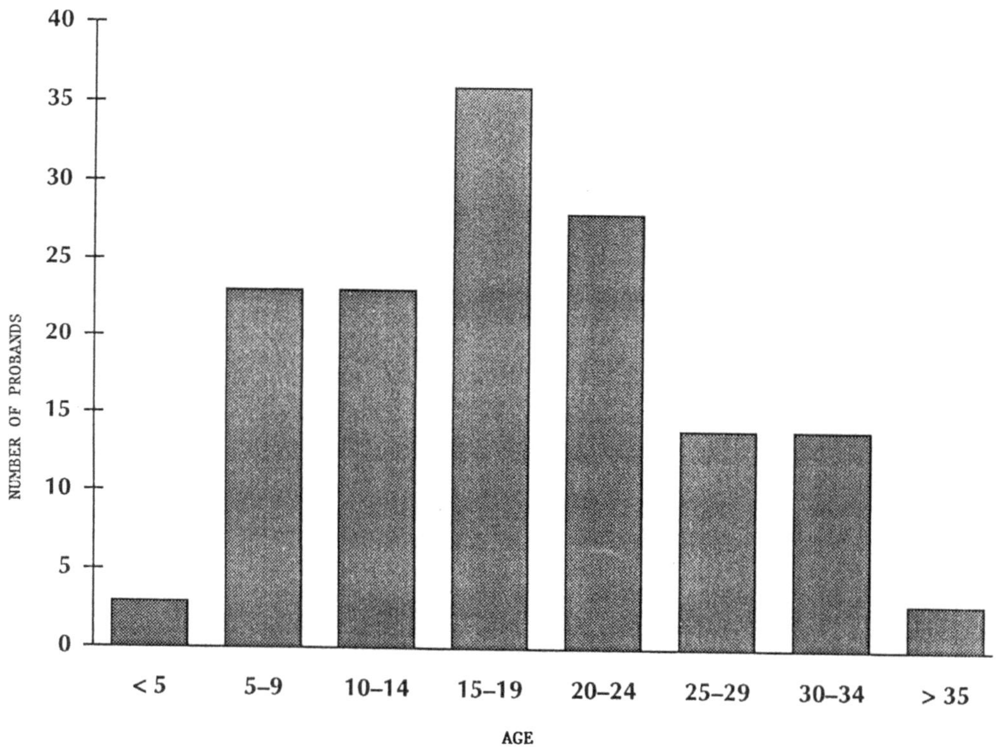

Figure 1. Ages of the probands

<u>Parents</u>: The parental age at the birth of the child with PWS was known for 137 mothers and 136 fathers (Figure 2). The mean age for mothers was 28.08 years ± SD 6.51 and the mean age for fathers was 30.45 years ± SD 7.10. These were not different from the median age of 28.3 years and 30.8 years, respectively, at the birth of all nuptial children in Australia in 1987 (Castle, 1987). Parental consanguinity was admitted in 2 families. In 2 families the mother required fertility medication to achieve pregnancy. Both of these women were short and obese (one of them (QM) is described in Table IV). In no other family was it difficult to achieve pregnancy. This 2/138 is statistically different (Chi^2 =

4.6, p<0.05) from the 1/8 couples in Australia who have fertility problems.

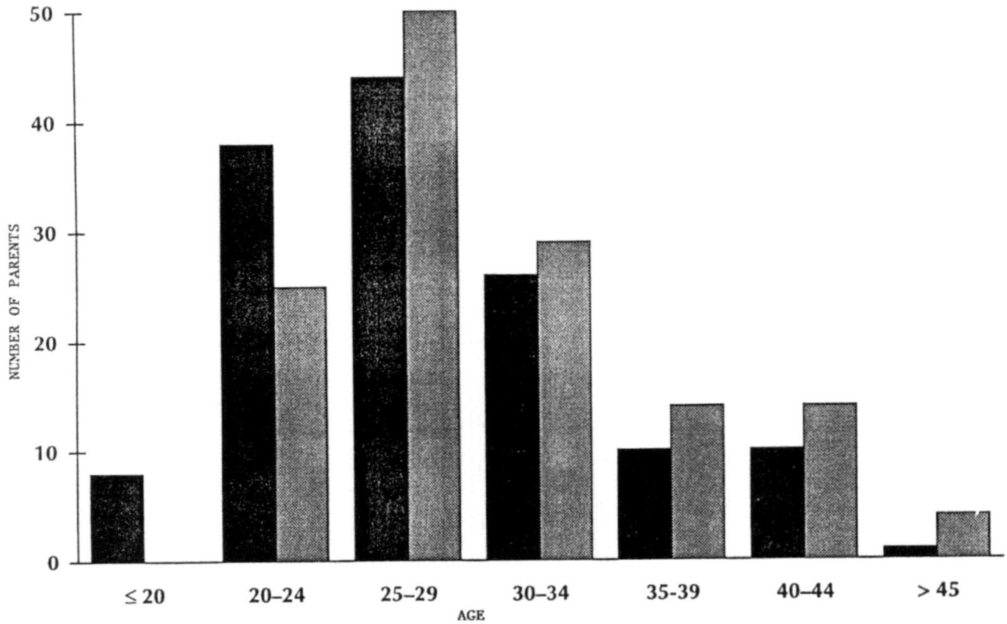

Figure 2. Ages of the mothers (left column) and fathers (right columnn) at the birth of the probands

Cytogenetic Findings: In 12 cases cytogenetics had not been performed. Of the remaining 132, high resolution banding (HRB) (Yunis, 1976) had been performed in 88 (66%). A cytogenetic summary is shown in Table I. The boy with 47,XXY had normal chromosome 15s and was phenotypically a classical PWS (Smith et al, 1990).

Siblings: There were 266 live born siblings in these PWSF, 133 males and 133 females. The age range of the siblings was 3 months to 42 years. The position in the family and the size of the sibship are shown in Table II. The adopted/fostered cases are counted as only children. There were 4 sets of twins (counted as one pregnancy) and this data are also shown in Table II. Thus PWS twins occurred in 1 in 36 families and 1 in 119 pregnancies (vide infra).

Table I. Cytogenetic Studies in Probands (N = 132)

88 Tested with HRB	- 45 with deletion 15q - 43 normal
1 Aneuploidy	- 47,XXY
5 Structural Rearrangements (3.8%)	

1 *de novo* duplication: 47,XY,+del(15)(pter-q12)
1 Robertsonian translocation: t(14q15q) (maternal) balanced
1 Robertsonian translocation: t(13q15q) (maternal) balanced
2 reciprocal translocation: 45,XX,t(7;15)(q32;q13) (unknown origin)
45,XY,t(9;15)(q34;q13) (*de novo*)

Table II. Sibling Data (N = 266)

133 Males, 133 Females

Size of Sibship	Position in Family
0 = 13	1st = 56
1 = 59	2nd = 47
2 = 47	3rd = 27
3 = 10	4th = 7
4 = 10	5th = 5
5 = 1	6th = 1
6 = 2	13th = 1
7 = 1	
15 = 1	
144	144

4 Sets of Twins

Family	Sex	Phenotype
1	MM	both PWS
1	FF	one normal; one PWS
1	MM	one normal; one PWS
1	FF	one PWS; one stillbirth

Pregnancy Loss: In this cohort there were 79 spontaneous pregnancy losses (Table III), consisting of 74 spontaneous abortions (SA), 4 stillbirths (SB) and one neonatal death (NND). Data from the two families with familial balanced translocations, which are known to adversely affect pregnancy outcome because of unbalanced segregants, have been excluded from this analysis. Thus total pregnancies for 142 PWSF is 479: 142 PWS, 258 living sibs and 79 pregnancy losses. The loss rate of recognized pregnancies is 16.5%. Recurrent loss (3 or more pregnancies) was seen in 6 families (bottom of Table III). This 6/142, or 4.2%, is greater than the Australian general population figure of 1 in 200 couples (McDonough and Tho, 1985), but does not reach statistical significance (Chi square = 2.5, p <0.2).

TABLE III - PREGNANCY WASTAGE

N	Type of Loss	# of Families		Total
1	SA	33	=	33
2	SA	6	=	12
3	SA	2	=	6
4	SA + NND	1	=	5
5	SA + SB	1	=	6
6	SA	1	=	6
8	SA	1	=	8
1	SB	3	=	3
				79

Recurrent Loss

N	Type of Loss	# of Families
3	SA	2
4	SA + NND	1
5	SA + SB	1
6	SA	1
8	SA	1
		6

Recurrence of PWS: None of 138 families had a recurrence of PWS within a sibship. In one family there was a recurrence of partial PWS in a second cousin. In this family the proband showed an interstitial deletion of chromosome 15. The male second cousin was cytogenetically normal on routine study. In Table IV are listed 5 PWSF in which a relative showed some features of PWS. Case 1 (QM) was evaluated by us (AS and JP), but the other relatives have not been seen by any of the authors. Complete family ascertainment of relatives was not available.

TABLE IV: Features of PWS in Family Members (5/138 Families)

1.	QM (Mother)	Obese 85kg (despite stapling operation); short; personality problem; history of infertility; short sighted; suggestive facies. normal intelligence; did not give early history of perinatal problems.
2.	RS (Maternal Aunt)	Small hands and feet; obese; short; slow learner at school.
3.	AS (2 Paternal Second Cousins)	Both had mental delay and obesity. Female born 1961. Female born 1938, died at 27 years. Consanguinity in this family.
4.	MA (Paternal Second Cousin)	Female with weight 133 kg at 21 years; died soon after; slow learner
5.	KT (Paternal Uncle)	Male born 1951; obese; short; mentally retarded

DISCUSSION

We have provided data on a cohort of Australian families with a PWS proband. The diagnosis of PWS was problematic in some cases, although not difficult in most classical older cases. Some authors consider the PWS phenotype straight forward at an early age, but this has not been our experience or that of others (Greenswag, 1987). Hence only cases over 3 years of age have been included and atypical cases have been specifically excluded. Over the time period of this data collection, which in some cases was prior to 1982 (Smith and Simpson, 1982), and considering the vast geographical spread of the cases, the cytogenetic analyses have yielded results similar to other studies (Butler, 1990).

The sex ratio of the probands was equal in this cohort of classical PWS (Greenswag, 1987) and confirms the ascertainment bias of young cases due to hypogenitalism. Parental ages at the birth of the PWS are not different from the Australian population data.

Normal parental age was also found by Holm and Sulzbacher (1987). Our conclusion is based on comparing parental ages with the 1987 population data (Castles, 1987). However, parental ages have been increasing in Australia. The mean maternal age of all births in 1970 was 26.3 years and mean paternal age was 29.8 years, while the means in 1989 were 28.2 years and 31.2 years, respectively. As 76.4% of the cases presented here were born before 1980, there is the suggestion of a slightly increased maternal age. This is important in evaluating maternal nondisjuntion as a likely causative mechanism in producing uniparental disomy. There is a suggestion of increased fertility in these PWSF, which has been commented on previously (Dunn, 1968). The pregnancy loss rate was 16.5% which is not different from the figure of 15% used in Australia as the population spontaneous abortion rate (Warburton and Fraser, 1964). The twinning rate was also normal (Wenstrom and Gall, 1988).

Our data supports the conclusion that PWS is sporadic. It does not provide support for autosomal recessive inheritance. There was no recurrence within a sibship and there was no increased consanguinity (Booy and Pearn, 1990). Recessive inheritance is quite unlikely, assuming a low gene frequency for a rare disorder. However, our data does not exclude the possibility of a small subgroup in which autosomal recessive inheritance is operative (Lubinsky et al, 1987; Burke et al, 1987). The findings of suggestive features of PWS in 5 cases is interesting, but given that obesity, mental retardation and short stature are all common, it cannot be concluded that this suggests incomplete expression of a PWS gene. Case I (QM) was the most convincing, but she was not fully worked up. Nevertheless, it appears that complete and thorough ascertainment of family members of PWSF is warranted to provide definitive data on this important issue of gene expression.

SUMMARY

A collaborative study of individuals with PWS families was undertaken to obtain a profile of these families in Australia.

Data were obtained through the PWS Association, PWS clinics, genetics clinics, Developmental Disability Service, pediatricians, general practitioners and other referrals in each state of Australia. There were 144 probands with classical PWS, 74 males and 70 females. Their ages ranged from 3.5 to 41 years. The mean ages of the parents at the birth of the child with PWS was 28.08 years (mothers) and 30.45 years (fathers). There were 266 siblings of these probands (133 males and 133 females), including 4 sets of twins. There was no recurrence within a sibship. Our data suggest that the families of individuals with PWS are normal and that PWS is a sporadic disorder.

REFERENCES

Booy R, Pearn JH (1990) Cousin Marriage. Human Genetics Society of Australia Bulletin Abstract. p 18

Burke CM, Kousseff BG, Gleeson M, O'Connell BM, Devlin JG (1987) Familial Prader-Willi syndrome. Arch Int Med 147:673-675

Butler MG (1990) Prader-Willi syndrome: Current understanding of cause and diagnosis. Am J Med Genet 35:319-332

Cassidy SB (1987) New data for determining Prader-Willi syndrome recurrence risk. Am J Med Genet 28:94

Castle I (1987) Australian Bureau of Statistics. Canberra. "In Births Australia 1987" Catalogue No. 3301.0 Commonwealth Government Printer. p 1-8

Clarren SK, Smith DW (1977) Prader-Willi syndrome. Am J Dis Child 131:798-800

Dunn HG (1968) The Prader-Labhart-Willi syndrome: Review of the literature and report of nine cases. Acta Paediat Scand Supplement 186:1-38

Greenswag LR (1987) Adults with Prader-Willi syndrome: A survey of 232 cases. Develop Med Child Neurol 29:145-152

Holm VA (1981) The diagnosis of Prader-Willi syndrome. In: Holm VA, Sulzbacher S, Pipes PL (eds). Prader-Willi Syndrome. University Park Press, Baltimore. p 27

Holm VA, Sulzbacher S (1987) Demography of Prader-Willi syndrome in the 1980s. Ninth Annual Meeting of Prader-Willi Syndrome Association, Houston

Ledbetter DH, Greenberg F. Holm VA, Cassidy SB (1987) Conference Report: Second Annual Prader-Willi Syndrome Scientific Conference. Am J Med Genet 28:779-790

Lubinsky M, Zellweger R, Greenswag LR, Larson G, Hansmarun I, Ledbetter DH (1987) Familial Prader-Willi syndrome with normal chromosomes. Am J Med Genet 28:37-43

McDonough PG, Tho SPT (1985) Aetiology and evaluation of repetitive reproductive wastage. In: Shearman RL (ed) Clinical Reproductive Endocrinology. Churchill Livingston p 205-210

Smith A, Simpson E (1982) Dermatoglyphic analyses of 24 individuals with the Prader-Willi syndrome. J Ment Defic Res 26:91-99

Smith A, Volpato F. Trent RJ (1990) Correlation of cytogenetic and DNA findings in Prader-Willi syndrome. In: Fraser WI (ed) Key Issues in Mental Retardation Research. Routledge Press p 57-63

Warburton D, Fraser FC (1964) Spontaneous abortion risks in man: Data from reproductive histories collected in a medical genetics unit. Am J Hum Genet 16:1-27

Wenstron KD, Gall SA (1988) Incidence, morbidity and mortality, and diagnosis of twin gestations. Clinics in Perinatology 15:1-12

Yunis JJ (1976) High resolution of human chromosomes. Science 191:1268-1270

A FAMILY FOCUSED CARE MODEL FOR PWS IN NORWAY - THE FRAMBU EXPERIENCE

Christian Aashamar
Frambu Health Centre
1404 Siggerud
Norway

Arvid Heiberg[1], Kari Storhaug[1] and Bjørn Lofterød[2]
Frambu Health Centre, Norway[1]; Moss Hospital, Norway[2]

Frambu Health Centre is an information and treatment center for persons with various types of handicapping conditions and their families. The Centre has evolved during more than 30 years from primarily taking care of patients with common disorders to taking care for patients with rare disorders and their families in recent years.

Norway has a decentralized National Health Care system which provides good medical services to the acutely ill. Families with chronically disabled members often complain about lack of expertise on the primary care level. This applies not only to medical but also to psychological, social, health insurance, technical aids, physiotherapy, dental and special education needs.

Frambu provides multiprofessional services during two week intensive courses for families with more than 25 rare diagnoses. It also provides summer camps for children.

All expenses for the families are covered, including expense for leave of absence for those professionally active. Most Norwegian families with Prader-Willi syndrome (PWS) members have been to Frambu in 1986, 1988 or 1991. The course program was planned and carried out in close cooperation with the Norwegian PWS Association and external medical experts on the condition.

NATO ASI Series, Vol. H 61
Prader-Willi Syndrome
Edited by S. B. Cassidy

The benefits for the families stem both from the expert information provided, and from learning from each other. They also provide their own collective experience for the external experts. On a long term basis this will result in better care.

The Norwegian PWS Organization was established at the first Frambu PWS course in 1986. More than 20 other small groups have also established lay organizations during their first Frambu course. The children and youth also benefit greatly when they meet others with the same problems and learn from them. They also obtain age appropriate information about their disorder. The sibs often attend the Centre with the disabled children, and all children are tutored by special educators (teachers/pre-school teachers).

All this leads to better family coping as contact, information and interaction is provided. The possibilities for research in the Frambu setting are very good, but underutilized. A survey of the Norwegian PWS families as well as a study of treatment with growth hormone have been initiated.

PRADER-WILLI SYNDROME IN NORWAY
An Epidemiological and Sociomedical Study

Bjørn Lofterød
Pediatric Outpatient Department
Moss Hospital
1500 Moss
Norway

Arvid Heiberg, Frambu Health Center, 1404 Siggerud, Norway

INTRODUCTION

In 1985 a survey of all patients with Prader-Willi syndrome (PWS) in Norway was performed. This survey was updated and supplemented in 1990. To our knowledge, this is the first national epidemiological study of PWS. Norway has 4.2 million inhabitants and is surrounded by the ocean on three sides. The country has common borders with Russia, Finland and Sweden. The population is stable and in our opinion ideal for epidemiological studies of this kind. The study also compares socioeconomic and living conditions with other well known handicaps such as Down syndrome, but this part is not yet completed and will be presented at a later date.

The study was done in close cooperation with Frambu Health Center. Frambu is a national information and treatment center for various types of handicapped subjects and their families. The Center has evolved over more than 30 years from taking care of patients with primarily common diseases to caring for patients with more rare disorders and their families. Frambu has two week courses for the handicapped individual and his/her family. All expenses for the families, including leave of absence from employment, are covered. The course program for PWS was planned and conducted together with the PWS Association and outside experts on the condition. Most of the Norwegian families with PWS members have been to Frambu. The benefits for the families are both the expert information provided and mutual support. The

NATO ASI Series, Vol. H 61
Prader-Willi Syndrome
Edited by S. B. Cassidy

PWS Association was formed in 1986 at the first Frambu course for the group.

The first patient with PWS in Norway was published as late as 1978 (Tangsrud, 1978). In 1981, the first deletion on chromosome 15 was described (Ledbetter et al, 1981). It was not until the PWS Association was formed that more professionals became interested in this very complicated syndrome. In 1989 one of the authors (Lofterod) attended PWS clinics in the USA at the University of Arizona and the University of Connecticut. In 1990 a 140 page book on the syndrome, written for parents and professionals was published in Norwegian.

METHODS

Questionnaires containing more than 200 questions were mailed to all parents of children with PWS in Norway. The questionnaire was divided into three parts. Part A compares social welfare, health and qualify of life of PWS individuals with non-handicapped people and their families. Part B compares same items with other well known handicaps such as Down syndrome. Part C contained specific questions on PWS including developmental history, physical characteristics, health problems and management.

RESULTS

A total of 71 patients with PWS were identified in Norway; all families were sent the questionnaire. The response rate was over 90% (Table I). Fifty-four of the 71 patients had been examined by the authors prior to the survey. After the study was completed, 68 patients had been examined and their diagnosis confirmed. Their ages ranged from 1 to 43 years of age.

Table I.

Number of Patients	71
Response rate	64/71 (90%)
Males	37 (52.1%)
Females	34 (47.9%)
Mean age	14 years
Age range	1-43 years
Number of patients examined by authors	54
3 patients had died in period 1985-1990	dates of birth (1961, 1964, 1987)

Table II shows that there were few patients in the older age groups. This probably represents a combination of lack of diagnosis and early death due to inadequate management. Table II also shows patients who were diagnosed by the authors after the survey was completed, but they are included in the incidence study (Table III). We believe that today no individual with PWS in Norway can start school without the diagnosis and without being referred to a specialist before the age of 15. We, therefore, think that our estimated incidence of 1 in 14,000 is reliable.

Table II.

Age Distribution

Date of Birth	Dec. '90	March '91
86 - 90	5	6
70 - 85	49	53
60 - 69	13	13
50 - 59	3	4
40 - 49	1	2
	71	78

Table III.

	Incidence
Children born 1970-1985 (N=53)	1/16,500
Children born 1974-1983 (N=39)	1/14,000

In spite of delayed diagnosis, 83% of the parents felt that something was wrong with their child before the age of 12 months. Ninety-two percent of the children had marked infantile muscular hypotonia and more than 50% required tube feeding for two weeks or more.

Hyperphagia began after 1 year of age, and 80% were hyperphagic before the age of 4. Dislike of drinking water was documented by parent report. As many as 76% of the parents informed us that their child disliked water. There seems to be a relationship between early diagnosis/good management and ability to accept drinking water.

Approximately 50% of the children never crawled before they learned to walk without support. It is well known that individuals with PWS, even as adults, show poor muscular coordination. More effort on early coordination exercises may reduce this problem.

The syndrome is a multihandicap disorder (Table IV). We found fewer patients with epileptic seizures than published in other studies (Holm, 1981). We postulate that other studies may have included some with somnolence as seizures of atypical absence. Other medical complications mentioned in Table IV are more in accordance with other studies.

Ninety-eight percent of the parents felt that they were manipulated by the behavior of their child. But only 46% reported that they received sufficient family support, compared to 98% of parents with non-handicapped children. This means that there is still a long way to go before the needs of a family with a PWS child are fulfilled in Norway. We hope that our study will make the social authorities in our country more aware of this fact.

Table IV.

Medical Complications

Diabetes mellitus	6.2%
Epileptic seizures	5.0%
Hypothyroidism (1/71)	1.4%
Congenital dislocation of hips (3/71)	4.2%
Strabismus	31%
Refractive errors	28%
Glasses	28%
No ophthalmologic problems	19%
Somnolence	82%

DISCUSSION

PWS is a multihandicap disorder. International cooperation is important for the understanding of the disease.

Good management depends on early diagnosis and professional team work. In Norway, considerable time is spent on providing information and assistance to the local support team, that is created in the community for every patient with PWS.

Our study shows many results that will be investigated further, and only a few are published in this article. But hyperphagia and obesity do still, in our opinion, represent the biggest problem for the patient and the family.

Most of the patients disliked water. We believe that this may contribute to obesity as water is usually replaced by a more calorie rich fluid. It may also be true that there is a relationship between poor muscular coordination and physical passivity. It so, more effort should be placed on coordination exercises early in childhood.

PWS is still a clinical diagnosis. Nearly all the patients

in this survey were examined by the authors. This, combined with the fact that there is a close cooperation among all the pediatric departments in Norway, make us believe that our estimated incidence is reliable.

REFERENCES

Holm VA (1981) The diagnosis of Prader-Willi syndrome. In Holm VA, Sulzbacher S, Pipes PL (eds): Prader-Willi syndrome. Baltimore, University Park Press, Chapter 3

Ledbetter DH, Riccardi VM, Airhart SD, Strobel RJ, Keenan SB, Crawford JD (1981) Deletions of chromosome 15 as a cause of the Prader-Willi syndrome. N Engl J Med 304:325-329

Tangsrud S (1978) Prader-Willi-Labhart syndrome. The Journal of the Norwegian Medical Association 16:839-840

Dr. Lofterød's new address is: Center of Habilitation, Sandveien 15, 1600 Fredrikstad, Norway

A MULTICENTER ITALIAN STUDY ON PRADER-WILLI SYNDROME

G.B. Pozzan
Department of Pediatrics
University of Padova
Via Giustiniani 3,
35128 Padova
Italy

F. Cerutti[2], A. Corrias[3], C. Maffeis[4], G. Tonini[5], M. De Simone[6], A. Criño[7], M. Sposito[8], A. Balsamo[9], P. Balestrazzi[10], T. De Toni[11], P. Brambilla[12], C. Livieri[13], A. Salvatoni[14], L. Vido[1], L. Pinelli[4], M.S. Lungarotti[8], L. Mazzanti[9], L. Iughetti[10], R. Gastaldi[11], R. Lorini[13], S. Seminara[15], P. Borrelli[7], G. Chiumello[12]

Department of Pediatrics, Padova, Italy[1]; Clinica Pediatrica Torino[2]; Divisione Endocrinologia Pediatrica Ospedale Regina Margherita Torino[3]; Clinica Pediatrica Verona[4]; Istituto per l'Infanzia Trieste[5]; Clinica Pediatrica L'Aquila[6]; Divisione Endocrinologia Ospedale Bambin Gesù Roma[7]; Clinica Pediatrica Perugia[8]; Clinica Pediatrica 2° Bologna[9]; Clinica Pediatrica Parma[10]; Clinica Pediatrica 2° Genova[11]; Clinica Pediatrica 3° H.S. Raffaele Milano[12]; Clinica Pediatrica Pavia[13]; Clinica Pediatrica Varese[14]; Clinica Pediatrica Firenze[15]

INTRODUCTION

Prader-Willi syndrome (PWS) is a relatively common disorder with an incidence estimated at about 1:10,000 (Holm, 1981) which represents the most common dysmorphic/genetic form of human obesity. The primary features of this condition include infantile hypotonia, failure to thrive, hypogonadism and developmental delay followed by development of obesity, short stature, mental retardation and behavior problems. Mild dysmorphism is part of the syndrome. About 59% of patients with PWS have an interstitial of the proximal long arm of chromosome 15 (Ledbetter et al, 1987).

We report on 77 patients with PWS collected through a multicenter study of the S.I.E.D.P. (Società Italiana di Endocrinologia e Diabetologia Pediatrica) obesity group. Forty-four of these patients were studied using high resolution cytogenetic analysis. Possible differences in clinical features between patients with interstitial deletion and those with a normal

NATO ASI Series, Vol. H 61
Prader-Willi Syndrome
Edited by S. B. Cassidy

karyotype were sought.

PATIENTS AND METHODS

Seventy-seven patients with the diagnosis of PWS (42 males and 35 females) were admitted to the study. The age range was 2.2 to 29.8 years. The criteria for the admission to the study were: 1) the presence of chromosome 15(q11-13) deletion associated with obesity or, in the absence of chromosome 15 deletion, the presence of at least five of the most important symptoms of PWS (neonatal hypotonia, low birth weight, obesity, small hands and feet, short stature, mental retardation, hypogonadism or cryptorchidism).
We collected anamnestic, auxological, clinical and metabolic data in five groups: 1) total, 2) with chromosome 15 deletion, 3) without chromosome 15 deletion, 4) males, and 5) females.

The auxological data (weight and height) were compared with Tanner growth charts. We considered "short stature" to be height lower than the 3rd centile and "obese" to be weight over 120% of ideal body weight for stature. "Hypogonadism" in male patients was defined as penis length <3 cm and/or cryptorchidism and/or bilateral testicular volume less than 4 ml after 14 years of age. "Hypogonadism" in female patients was defined as no breast development at age 14 and/or primary amenorrhea after 15 years of age. Hand and foot lengths were compared respectively with Feingold and Blais standards: a hand or a foot was considered "short" if below the 3rd centile.

Metabolic assessment was made by determining the prevalence of diabetes mellitus and impaired glucose tolerance (according to the American Diabetes Association criteria) as well as by measuring the blood levels of glycosylated hemoglobin, fructosamine, cholesterol, HDL cholesterol and triglycerides.

Mean, SD, Chi square and student's t tests were used for statistical analysis.

RESULTS

Survival: Three patients (1 girl and 2 boys) died at the age of 3, 2 and 17 years of pneumonia, cardiac failure and cardiac infarction, respectively.

Karyotype: Eighteen patients were investigated with standard technique, and all showed a normal karyotype. Forty-four patients were investigated with high resolution technique. In 23 (52%) the karyotype results were normal, in 18 (41%) deletion of 15q12 was present. In three female patients (7%) other chromosome abnormalities were demonstrated: 46,XX,t(13-14); 46,XX,inv(9)(p11q12); 46,XX,t(15q;18q).

Clinical Data: The results of anamnestic, auxological, clinical and metabolic data of patients in the deletion group was compared with those in the chromosomally normal group (Table I). Short stature was less frequent in patients with chromosome 15 deletion ($p \leq 0.05$) and hypogonadism/cryptorchidism was more frequent in the same group ($p \leq 0.05$). No significant differences were found between males and females except for the greater incidence of hypogonadism ($p \leq 0.05$) (Table II).

Table I: Clinical Features of 77 Patients with PWS and Comparison Between Groups With and Without Deletion of Chromosome 15

	Total	With Deletion	Without Deletion	X^2/t(p)
n°	(77)	(18)	(23)	
age (years) (mean ± SD)	12.2±5.8	13.0±6.0	13.6±7.2	NS
age at diagnosis (years) (mean ± SD)	6.0±4.5	6.3±5.0	7.2±5.7	NS
age at onset of obesity (years) (mean ± SD)	2.8±2.2	3.0±3.0	2.5±1.4	NS

birth weight (gm) (mean ± SD)	2718±478	2728±373	2769±573	NS
birth weight <2500 gm	25/77 (32%)	6/18 (33%)	7/23 (30%)	NS
abnormal delivery	40/75 (53%)	11/17 (65%)	8/21 (38%)	NS
neonatal hypotonia	70/74 (95%)	17/18 (94%)	21/22 (96%)	NS
obesity	76/77 (99%)	18/18 (100%)	23/23 (100%)	NS
weight (% IBW) (mean ± SD)	174±44	159±42	178±39	NS
stature <3rd centile	44/76 (58%)	7/17 (39%)	17/23 (74%)	<0.05
Z score for height (mean ± SD)	-2.05±1.6	-1.49±1.45	-2.20±1.38	NS
small hands and feet	69/73 (95%)	18/18 (100%)	21/22 (96%)	NS
hypogonadism or cryptorchidism	45/58 (78%)	12/13 (92%)	10/17 (58%)	<0.05
hyperphagia	53/61 (86%)	13/14 (92%)	15/19 (79%)	NS
sleep apnea	17/42 (41%)	5/10 (50%)	5/13 (38%)	NS
ocular disturbances*	39/61 (64%)	9/13 (69%)	14/17 (82%)	NS
mental retardation	73/75 (97%)	16/17 (94%)	23/23 (100%)	NS
scoliosis	42/72 (59%)	12/16 (75%)	11/21 (52%)	NS
orthopedic problems**	64/72 (88%)	18/18 (100%)	21/21 (100%)	NS
diabetes***	4/61 (6%)	3/14 (21%)	1/18 (5%)	NS
impaired glucose tolerance	4/61 (6%)	1/14 (7%)	2/18 (11%)	NS
normal glucose tolerance	53/61 (86%)	10/14 (71%)	15/18 (83%)	NS
HbA1c (%) (mean ± SD)	5.6±1.4	5.9±1.4	6.2±1.7	NS

fructosamine (mMol/l) (mean ± SD)	2.2±0.3	2.2±0.3	2.2±0.3	NS
cholesterol (mg/dl)**** (mean ± SD)	192±36	192±42	183±37	NS
HDL chol.(%) (mean ± SD)	50.2±21.3	51.0±24.0	52.0±30.0	NS
triglycerides (mg/dl)***** (mean ± SD)	95.7±47.3	107.0±56.0	91.0±48.0	NS
hypo-thyroidism	1/77	0/17	1/23	

IBW = Ideal body weight. See also Table II legend.

Table II: Clinical Features of 77 Patients with PWS and Comparison Between Male and Female Groups

	Total	Males	Females	X^2/t(p)
n°	77	42	35	
age (years) (mean ± SD)	12.2±5.8	12.2±5.8	12.2±5.8	NS
age at diagnosis (years) (mean ± SD)	6.0±4.5	5.5±4.1	6.7±4.9	NS
age at onset of obesity (years) (mean ± SD)	2.8±2.2	3.1±2.2	2.4±2.1	NS
birth weight (gm) (mean ± SD)	2718±478	2668±370	2783±543	NS
birth weight <2500 gm	25/77 (32%)	15/42 (35%)	10/25 (28%)	NS
abnormal delivery	40/75 (53%)	25/40 (62%)	15/35 (42%)	NS
neonatal hypotonia	70/74 (95%)	39/40 (98%)	31/34 (91%)	NS
obesity	76/77 (99%)	42/42 (100%)	34/33 (97%)	NS

weight (% IBW) (mean ± SD)	174±44	182±49	168±32	NS
stature <3rd centile	44/76 (58%)	23/41 (56%)	21/35 (60%)	<0.05
Z score for height (mean ± SD)	-2.05±1.6	-2.18±1.7	-1.90±1.4	NS
small hands and feet	69/73 (95%)	38/40 (95%)	31/33 (93%)	NS
hypogonadism or cryptorchidism	45/58 (78%)	39/41 (95%)	6/17 (35%)	<0.05
hyperphagia	53/61 (86%)	31/35 (88%)	22/27 (81%)	NS
sleep apnea	17/42 (41%)	10/22 (45%)	7/20 (35%)	NS
ocular disturbances*	39/61 (64%)	19/35 (54%)	20/26 (76%)	NS
mental retardation	73/75 (97%)	39/40 (97%)	34/35 (97%)	NS
scoliosis	42/72 (59%)	21/38 (55%)	21/33 (63%)	NS
orthopedic problems**	64/72 (88%)	36/39 (92%)	28/33 (84%)	NS
diabetes***	4/61 (6%)	2/33 (6%)	2/28 (7%)	NS
impaired glucose tolerance	4/61 (6%)	2/33 (6%)	2/28 (7%)	NS
normal glucose tolerance	53/61 (86%)	29/35 (82%)	24/28 (85%)	NS
HbA1c (%) (mean ± SD)	5.6±1.4	5.5±0.9	5.7±1.7	NS
fructosamine (mMol/l) (mean ± SD)	2.2±0.3	2.2±0.3	2.2±0.3	NS
cholesterol (mg/dl)**** (mean ± SD)	192±36	193±35	189±35	NS
HDL chol.(%) (mean ± SD)	50.2±21.3	49.0±23.0	50.0±18.0	NS
triglycerides (mg/dl)***** (mean ± SD)	95.7±47.3	101.0±52.0	88.0±39.0	NS

hypo-thyroidism	1/77	0/42	1/35

*ocular disturbance = 54% myopia, 16% strabismus, 13% disorder of pigmentation, 11% astigmatism, 5% hyperopia
**orthopedic problems = 86% genu valgum, 3% osteoporosis, 3% joint hypermobility, 7% others
***two patients with diabetes are treated with insulin and 2 with antihyperglycemic drugs
****the cholesterol level was above 200 mg/dl in 26/67 (39%) and above 250 in 3/67 (4.5%).
*****the triglyceride level was above 150 mg/dl in 9/63 (14%) and above 200 mg/dl in 4/63 (6.3%).

COMMENTS

Prader-Willi syndrome is a relatively common disorder with an incidence estimated at about 1:10,000 (Holm, 1981) without racial differences. In spite of this incidence, only 77 patients were referred to us from the main Italian infantile endocrinological services in the north and center of our region. PWS is therefore probably still underdiagnosed in Italy. Subjects with this problem are probably referred to health services different from pediatric endocrinology, and consequently are not included in this study.

The prevalence of evident deletion of 15q12 of in 44 children who had high resolution karyotype was about 40%, slightly lower than in other studies which averaged 59% with this chromosomeal alteration in PWS (Ledbetter et al, 1987), but it is very close to the data published in a similar study by Wenger et al (1987).

Fear et al (1985), Takano et al (1986) and Wenger et al (1987) did not demonstrate any clinical differences between PWS patients with evident deletion and those without it. In our series, short stature is less frequent and hypogonadism more frequent in patients with chromosome 15 deletion. Also, the weight excess appears smaller in patients with chromosome deletion, but this difference was not statistically significant.

There were no physical or metabolic differences between male and female groups. The only difference between males and females was the prevalence of hypogonadism, but this is probably due to the difficulty of diagnosing hypogonadism in the female in the prepubertal period while cryptorchidism is evident even in the male infant. Others have reported clinical differences between patients with deletion and without deletion. Mattei et al (1983) noted the absence of feeding difficulties and a lower incidence of hypotonia in cases with a normal karyotype; Wiesner et al (1987) and Butler (1989) reported that patients with 15(q11-13) deletion had lighter hair color, more sun sensitivity and fairer complexion than those without deletion. Our data do not confirm that of Mattei et al, and data concerning hair and skin pigmentation in our patients are not at present available.

It must be pointed out that only 44 of our 77 patients were tested with high resolution karyotype study, and the proportion of deleted patients in our entire population is unknown.

Finally, the high incidence of death at a young age in our series was remarkable. One boy died at the age of 17 years of cardiac infarction. Atherosclerotic coronary artery disease has been documented in PWS (Reed et al, 1968; Lamb and Johnson, 1987), confirming the potential role of coronary pathology as a cause of mortality in these patients. For this reason, and also because of the prevalence of other complications of obesity, such as orthopedic problems (especially genu valgum and scoliosis) and non-insulin dependent-diabetes, we recommend an earlier diagnosis of PWS. In Italy the mean age at diagnosis is 6 years, about three years after the onset of obesity, when medical and behavioral approaches to obesity and prevention of metabolic complications are more difficult.

It is thus suggested that every infant with neonatal hypotonia or feeding problems in the first year of life, be evaluated for PWS without waiting for other features such as obesity, short stature and hypogonadism which are late in onset and sometimes absent, as is demonstrated by our data.

REFERENCES

Butler MG (1989) Hypopigmentation: A common feature of Prader-Labhart-Willi syndrome. Am J Hum Genet 45:140-146

Fear CN, Mutton DE, Berry AC, Heckmatt JZ, Dubowitz (1985) Chromosome 15 in Prader-Willi syndrome. Develop Med Child Neurol 27:305-311

Holm VA (1981) The diagnosis of Prader-Willi syndrome. In Holm VA, Shulzbacher S, Pipes PL (eds) Prader-Willi syndrome, University Park Press, Baltimore, p 27

Lamb AS, Johnson WM (1987) Premature coronary artery atherosclerosis in a patient with Prader-Willi syndrome. Am J Med Genet 28:873-880

Ledbetter DH, Greenberg F, Holm VA, Cassidy SB (1987) Conference report: second annual Prader-Willi scientific conference. Am J Hum Genet 28:779-790

Mattei JF, Mattei MG, Giraud F (1983) Prader-Willi syndrome and chromosome 15. Hum Genet 64:356-362

Reed WB, Ragsdale W, Curtis AC, Richards HJ (1968) Acanthosis nigricans in association with various genodermatoses. Acta Derm-Venereol 48:465-473

Takano T, Nakagome Y, Nagafuchi S, Tanaka F, Nakamura J, Nagano T, Tanae A, Hibi I (1986) High resolution cytogenetic studies in patients with Prader-Willi syndrome. Clin Genet 30:241-248

Wenger SL, Hanchett JM, Steel MW, Maier BW, Golden WL (1987) Clinical comparison of 59 Prader-Willi patients with and without the 15 (q12) deletion. Am J Med Genet 28:881-887

Wiesner GL, Bendel CM, Olds DP, White JG, Arthur DC, Ball DW, King RA (1987) Hypopigmentation in the Prader-Willi syndrome. Am J Hum Genet 40:431-442

A COMPARISON OF CHARACTERISTICS IN 33 JAPANESE AND 83 AMERICAN PATIENTS WITH PRADER-WILLI SYNDROME

Jeanne M. Hanchett
The Rehabilitation Institute of Pittsburgh
6301 Northumberland Street
Pittsburgh, Pennsylvania 15217
USA

Nobutake Matsuo[1], Toshiro Nagai[1], Norio Niikawa[2], and Hidefumi Tonoki[3]

Department of Pediatrics, Keio University School of Medicine, Tokyo, Japan[1]; Department of Human Genetics, Nagasaki University School of Medicine, Nagasaki, Japan[2]; Department of Pediatrics, Hokkaido University School of Medicine, Sapporo, Japan[3]

INTRODUCTION

Prader-Willi syndrome (PWS) was identified in 1956 (Prader et al, 1956) and has been reported to occur in all racial groups (Golden et al, 1984; Ikeda et al, 1973). Clinical diagnosis is based on the presence of hypotonia in infancy, obesity occurring after age 2, hypogonadism (which is evidenced by cryptorchidism in males), and mild mental retardation. Additional findings include skin picking, speech problems, behavior problems, and short stature.

In order to ascertain whether clinical findings differ in racial and cultural groups, the above 8 characteristics of PWS were compared between 33 Japanese patients and 83 American patients, including 77 caucasian patients and 6 black patients.

METHODS

The average age of Japanese patients was 9.5 years with a range of 8/12 to 23 10/12 years; average age of American patients was 13.3 years with a range of 3 10/12 to 22 3/12 years. Twenty-

NATO ASI Series, Vol. H 61
Prader-Willi Syndrome
Edited by S. B. Cassidy

four of the Japanese patients were males; forty of the American patients were males. Patients examined in Japan were outpatients seen in 3 medical centers. Patients at The Rehabilitation Institute of Pittsburgh in the United States were inpatients and had been referred for management from all parts of the United States. Height and weight measurements for American children were compared to standard growth charts as determined by the National Center for Health Statistics. Japanese children's measurements were compared to norms established by one of us (NM) (Tsuzaki et al, 1987). Statistical significance was determined using the chi square method.

RESULTS

Differences were found in the amount of obesity; Japanese patients averaged 50.6% overweight with a range of 17% underweight to 134% overweight. American patients averaged 85.7% overweight with a range of 0%-300% overweight. The 6 black patients averaged 125% overweight. If patients over the age of 3 years were compared, the percentage overweight in Japanese children increased to 60.2%. These two differences in weight are significant at the $p = .001$ level and $p = .01$ level, respectively. The 5 Japanese patients who were under age 3 were within 20% of ideal body weight, as one would expect in this syndrome. No American patients were under 3 years of age. Twenty-two of 28 Japanese patients were more than 20% overweight (78.5%) compared to 74 of 83 American patients (89.2%). The remainder of patients, 6 Japanese (21.4%) and 9 Americans (10.8%), were within 20% of ideal body weight. That is, twice as many Japanese patients over 3 years of age were within 20% of ideal body weight; however, this did not reach statistical significance. Morbid obesity (defined as 100% or more overweight) occurred twice as often in American patients (41.0%) as in Japanese patients (21.4%).

Differences also occurred in skin picking, speech problems and behavior problems. Only 12% of Japanese patients picked their skin as compared to 73% of American patients. Speech problems were identified in 62.5% of Japanese patients, in 95% of Americans. Behavior problems were reported in 95% of Americans and only 36% of Japanese patients. All of these are significant at the p = .001 level.

No significant differences were found in 4 characteristics: early hypotonia, developmental delay, cryptorchidism in males and short stature. Early hypotonia and developmental delay occurred in virtually all Japanese and American patients; hypotonia occurred at 97% and 96%, respectively, and developmental delay occurred in 100% and 98.8%, respectively. Hypogonadism as evidenced by cryptorchidism in males was identified in 83% of Japanese males (20 of 24 patients). It was found in 97.5% of American males (39 of 40 patients). Short stature (defined as less than the 3rd centile on Japanese charts and less than the 5th centile on American charts) was found in 42% of Japanese patients and 58% of American patients. Since short stature may not be present in young patients with PWS, patients were divided into those under 10 years of age and over 10 years of age. There was no significant difference between the Japanese and American patients. Young patients in both groups were much less likely to have short stature (21% of Japanese, 29% of American) than older children (71% in Japanese, 65% in American).

Thus, significant differences were found between Japanese and American patients with PWS in obesity, skin picking, speech problems, and behavior problems, but not in early hypotonia, developmental delay, cryptorchidism, and short stature.

DISCUSSION

We propose the following possible explanations for the observed differences. Obesity in the general population is less common among Japanese; it is estimated to occur in 3% to 7% (Statistical report of health in school children, 1990) of Japanese school children as compared to 15% to 30% of American children (Wilson, 1987; Pipes, 1989). This is probably related to differences in types of food served in Japanese homes and American homes; the typical American diet is higher in fats and lower in carbohydrates than the typical Japanese diet. Cultural attitudes towards the importance of conforming to social practices may also play a role in controlling food intake in Japanese children. Japanese children may have more physical exercise than Americans. Referral patterns in the two countries may be different; Japanese patients were usually seen as outpatients as compared to American patients who were evaluated as inpatients. These American patients often had marked obesity and may not be representative of all patients with PWS in the United States.

Similar reasons may also explain differences in the frequency of behavior problems. Structured settings in the United States have been shown to decrease behavior problems in many patients with PWS (Thompson et al, 1988). Therefore, the structure of Japanese family life is undoubtedly helpful to children with PWS. Differences in the occurrence of skin picking and speech problems are difficult to explain and deserve further exploration.

It is clear that the basic findings of PWS are similar in both racial-cultural groups. Other characteristics of this syndrome do not appear to be found consistently in American and Japanese cultures.

REFERENCES

Golden WL, Hanchett JM, Breslin M, Steele MW (1984) Prader-Willi syndrome in black females. Clin Genet 26:161-163

Ikeda K, Asaka A, Inouye E, Kaihara H, and Kinoshita K (1973) Monozygotic twins concordant for Prader-Willi syndrome. Jap J of Hum Genet 18:220-225

Pipes PL (1989) Special concerns of dietary intake during infancy and childhood. In: Pipes Rl (ed) Nutrition in infancy and childhood. Times Mirror, CV Mosby, St. Louis, p 1381

Prader A, Labhart A, and Willi H (1956) Ein Syndrome von Adipositas, Kleinwuchs, Kryptochismus und Oligophrenie nach myatonieartigem Zustand in Neugeborenenalter. Schweiz Med Wochenschr 86:1260-1261

Statistical report of health in school children. MESC3-9101, (eds) Education Ministry. (Japan) 1990. p 130-131

Thompson D, Greenswag L, and Eleazer R (1988) Residential programs for individuals with Prader-Willi syndrome. In: Greenswag R and Alexander RC (eds) Management of Prader-Willi Syndrome. Springer-Verlag, New York, p 205

Tsuzaki S, Matsuo M, Osano M (1987) The physical growth of Japanese children from birth to 18 years of age. Helv Paed Acta 42:111-119

Wilson MH (1987) Obesity In: Hoekelman RA (ed) Primary Pediatric Care. CV Mosby, St. Louis, p 1381

ENDOCRINE PHYSIOLOGY AND THERAPY IN PRADER-WILLI SYNDROME

E. Martin Ritzén, M.D.
Pediatric Endocrinology Unit
Karolinska Hospital
104 01 Stockholm
Sweden

Per Bolme[2] and Kerstin Hall[1]

Departments of Endocrinology, Karolinska Hospital[1], and Department of Pediatrics, Huddinge Hospital[2]; Stockholm, Sweden

INTRODUCTION

Many of the symptoms that make up the complex clinical picture of Prader-Willi syndrome (PWS) point at derangements in hypothalamic function: short stature, hypogonadism, loss of appetite control, diminished pain sensitivity, excess sleepiness. Most of these symptoms were described in the original report of nine patients by Prader et al (1956), and confirmed by others (Jeffcoate et al, 1980; Bray et al, 1983). However, it has become evident that diabetes mellitus is not as common as was originally thought, and it has been argued that many of the endocrine problems are merely secondary to obesity. This chapter will give a brief review of the endocrine manifestations of PWS, against the background of our own comparison of PWS patients with a group of children and adolescents with "simple" obesity.

PATIENTS AND PROCEDURES

The patients are described in Tables I and II. They were all referred to the Departments of Pediatrics at the Karolinska or Huddinge Hospitals in 1978-1981. The protocol on admission included one week of laboratory investigations during which time the patients were allowed to eat the regular hospital meals.

The endocrine work-up included an arginine-insulin tolerance

NATO ASI Series, Vol. H 61
Prader-Willi Syndrome
Edited by S. B. Cassidy

test (only in PWS patients), a combined i.v. gonadotrophin releasing hormone/adrenocorticotropic hormone/thyroid releasing hormone (GnRH/ACTH/TRH) test (100 μg GnRH, 0.25 mg tetracosactide, 200 μg TRH per 1.73 m^2 body surface) and an overnight blood sampling every four hours. All blood samples were obtained after overnight fasting. A glucose infusion tolerance test (Grill and Efendic, 1987) was performed by a bolus injection of glucose, calculated to give a blood glucose concentration of 18-20 mmol/l followed by a 1 hour infusion of glucose to maintain the blood concentration at about 20 mmol/l. The hormones were measured by standard radioimmunoassay (RIA) techniques. Insulin-like growth factor-1 (IGF-1) was determined by both radioimmunoassay and radioligand methods (Hall et al, 1980). Insulin-like growth factor binding protein-1 (IGFBP-1) was determined by radioimmunoassay (Povoa et al, 1984). Melatonin measurements were kindly done by Dr. L. Wetterberg, Department of Psychiatry, St. Goran's Hospital, Stockholm.

The study was approved by the Karolinska Institute Ethics Committee.

RESULTS

Growth

The control obese patients were of average height but markedly overweight (+4.2 SDS, Table I). The short stature of the PWS patients became more pronounced with advancing age; the three children between 7 and 10 years of age had a mean height of -1.4 SDS, while the four above 16 years averaged -3.8 SDS. The same was true for overweight; while the body mass index of the youngest three averaged 25.7 (± SD 4.0), it was 43.3 (± SD 13.2) for the four above 16 years.

Evaluation of growth hormone secretory capacity was done in three different ways; in blood samples drawn before and one hour after falling asleep, before and after 20 minutes of exercise on an ergometer bicycle and (in PWS patients only) during an arginine-insulin tolerance test (AITT).

Table I. Patients under study. In the PWS group, there are 3 males and 5 females, in the controls 6 males and 4 females. Body mass index is calculated (weight in kg and height in meter) from the formula weight/height^2

	PWS (n=8) Mean (range)	Obese controls (n=10) Mean (range)
Age (yr)	13.9 (7.6-19)	10.2 (7.7-12.8)
Bone age (yr)	12.8 (6.5-17)	10.6 (8.1-13)
Height (cm)	137.1 (116-151)	142.9 (129.3-165)
Height SDS	-2.8 (-0.25--5.3)	+1.06 (-1.3 to +1.5)
Weight (kg)	67.3 (30-125)	61.9 (41.2-82.8)
Weight SDS	+2.5 (-0.6-+5.9)	+4.17 (+2.3 to +6.0)
Body mass index	34.3 (22-58)	29.2 (24-39)

The results are summarized in Table III. There were large variations between individuals in both groups. Although the mean growth hormone (GH) levels were lower in PWS, these reached statistical significance ($p<0.05$) only in samples drawn one hour after falling asleep. Only two out of seven PWS patients reached a maximum level of 10 µg/l or more during AITT. IGF-1 measured by radioreceptor assay (but not by RIA) was lower in the PWS group. However, the serum levels of immunoreactive IGF-1 were low, 34% and 45% of means in age-matched healthy children in the PWS and obese groups, respectively. The IGFBP-1 levels, which reflect the insulin production, did not differ between the two groups.

Hypothalmic-pituitary-gonadal axis

Since there were children and adolescents of both

prepubertal and pubertal age in the PWS group, the results cannot be readily summarized and compared with the prepubertal controls. However, none of the five PWS individuals who were more than 14 years of age had reached an adequate pubertal development for age, except for pubic hair; this was regularly well developed (Tanner stage III-IV) in all adolescent PWS patients, even if testes of the boys were small, and breast tissue was nonpalpable in the girls. Thus, adrenarache preceded "gonadarche". One PWS girl (18 years) had sparse and infrequent menstruations since 16 1/2 years of age, while two 16- and 18-year old girls had no evident signs of estrogen stimulation. Breast glands were difficult to evaluate because of obesity and lipomastia. Testosterone and estradiol levels in blood were subnormal for age. The pubertal status and baseline testosterone/estradiol levels are presented in Table II.

An acute challenge test with i.v. gonadotrophin releasing hormone (GnRH) caused a good leutinizing hormone (LH) release in three 16-17-year old PWS patients, but no detectable LH or follicle stimulating hormone (FSH) (<2 IU/l) in a 19-year old girl (H.E.) (Figure 1). In spite of this, her estradiol level in blood was 280 pmol/l. She was also the most obese, weighing 125kg at a height of 146.5cm (body mass index = 59).

Pituitary-adrenal axis

It was remarkable that both groups showed the lowest blood cortisol levels at 20.00 hours, rather than the expected 02.00 (Figure 2). However, there was no difference between the groups, and both responded equally (but somewhat delayed) to a bolus dose of ACTH (Figure 3). Unfortunately, dihydroepiandrostenedione (DHEA) and androstenedione were not measured.

Pituitary-thyroid axis

Although blood levels of triiodothyronine (PWS 2.03 +/-

Table II. Pubertal development and sex hormone levels in PWS and control obese patients

PWS patients

Initials	Sex M/F	Age yr	Pubic hair stage	Testes vol. ml	Breast /Genit stage	Testo-sterone nmol/l	Estra-diol pmol/l	Menarche age yr
M.I.	M	9.8	2	1	1	<0.3	-	
J.L.	M	14.6	3	2	1	4.3	-	
M.K.	M	17.9	4	5-8	3	5.5	-	
K.J.	F	7.0	1	-	1	-	81	-
E.L.	F	8.8	1	-	1	-	90	-
K.S.	F	16.1	3	-	3	-	-	-
L.J.P.	F	17.1	5	-	5	-	153	16.5
H.E.	F	18.0	3	-	?	-	280	-

Obese controls

Initials	Sex M/F	Age yr	Pubic hair stage	Testes vol. ml	Breast /Genit stage	Testo-sterone nmol/l	Estra-diol pmol/l	Menarche age yr
E.P.	M	8.5	1	1-2	1		-	
M.K.	M	8.5	1	2	1	<0.4	-	
N.F	M	10.3	1	1	1	1.7	-	
A.C.	M	10.3	1	1	1	0.6	-	
R.B.	M	11.9	1	1-2	1	0.2	-	
P.L.	M	12.3	1	1-2	1	1.1	-	
A.L.	M	12.9	1	*	1	1.7	-	
E.T.	F	8.30	1	-	1	-	55	-
J.B.	F	8.3	1	-	1	-	58	-
K.R.F.	F	9.0	1	-	1	-	31	-

* Bilaterally cryptorchid

Table III. Growth hormone (GH) and IGF-1 levels in blood from obese control and PWS patients. Mean ± s.e.m. IGF-1 was measured by two methods: Radioimmunoassay (RIA) and by radioreceptor assay (RRA)

	Obese controls (n)	PWS (n)
GH before sleep (μg/l)	1.12 ± 0.54 (7)	0.27 ± 0.04 (7)
After 1 h sleep	4.06 ± 1.31 (7)	0.51* ± 0.10 (7)
Max GH during exercise	5.67 ± 1.41 (7)	3.52 ± 2.19 (6)
Max GH during AITT	-	15.9 ± 4.9 (7)
IGF-1 (RIA, units/ml)	0.53 ± 0.06 (10)	0.52 ± 0.09 (6)
IGF-1 (RRA, units/ml)	0.78 ± 0.11 (10)	0.49 ± 0.05 (6)
IGFBP-1 (μg/l)	34 ± 8 (10)	33 ± 8 (7)

* $p < 0.05$

0.37, obese controls 2.35 +/- 0.53 nmol/l), thyroxine (PWS 93 +/- 21, obese controls 108 +/- 25 nmol/l, mean +/- S.D.) and baseline thyroid stimulating hormone (TSH) were not significantly different between the two groups, TSH (but not prolactin) release after a bolus dose of TSH was higher in the PWS patients (Figures 4 and 5). All the patients were clinically euthyroid.

In contrast to the perturbed cortisol rhythm, melatonin levels in blood showed the expected high levels at 02.00 hours low at 08.00 and 20.00 hours (Figure 6).

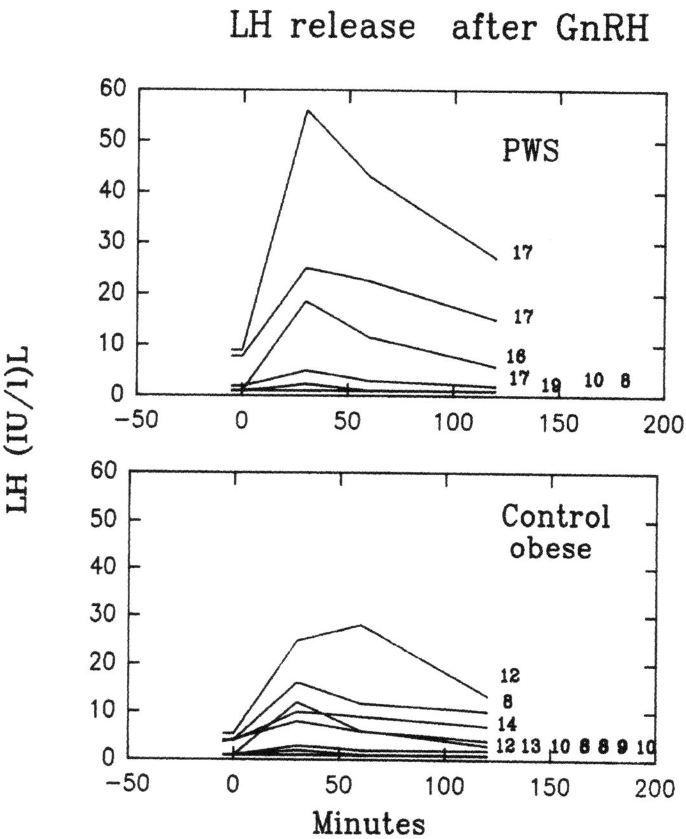

Figure 1. Blood levels of LH and FSH following an i.v. bolus dose of GnRH (100 µg/1.73 m^2 body surface). The age of the patients is indicated in the figure.

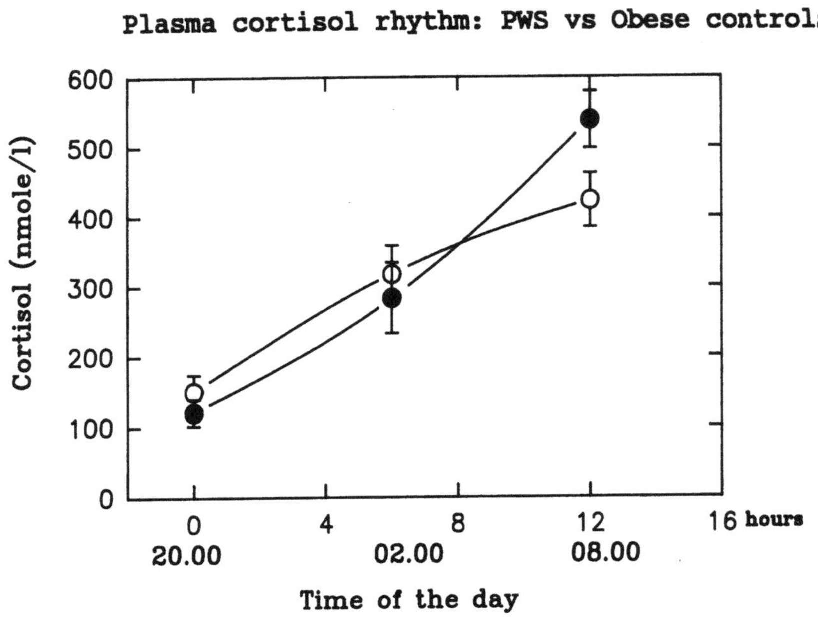

Figure 2. Cortisol levels in blood at three different times of the night in eight PWS (o-o) and ten control obese (●-●) patients.

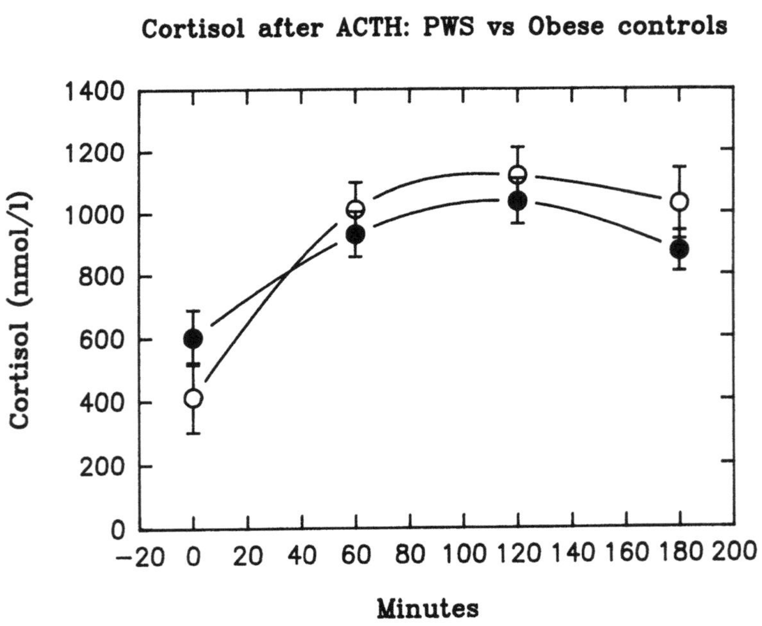

Figure 3. Cortisol levels in blood from PWS (o-o) and control obese (●-●) patients, following a bolus dose of synthetic $ACTH_{1-29}$.

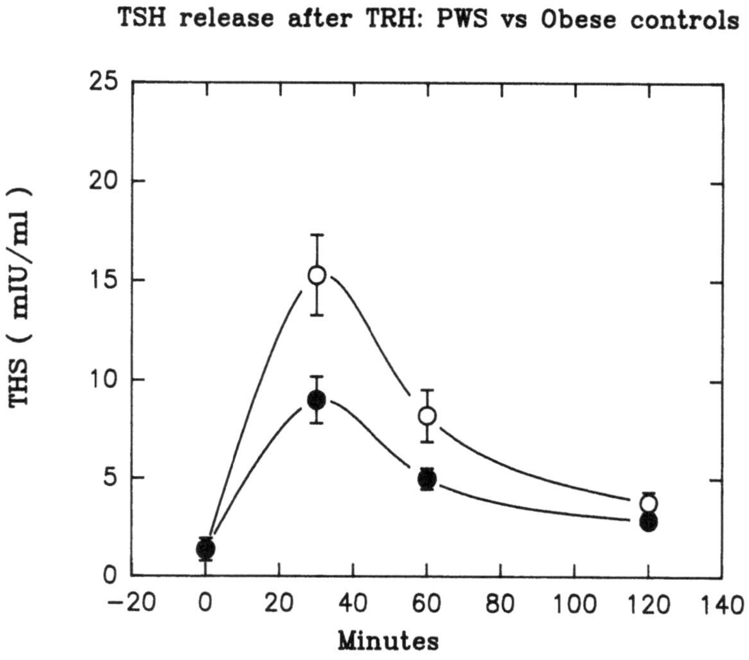

Figure 4. TSH levels in blood from PWS (o-o) and control obese (●-●) patients, following a bolus dose of TRH.

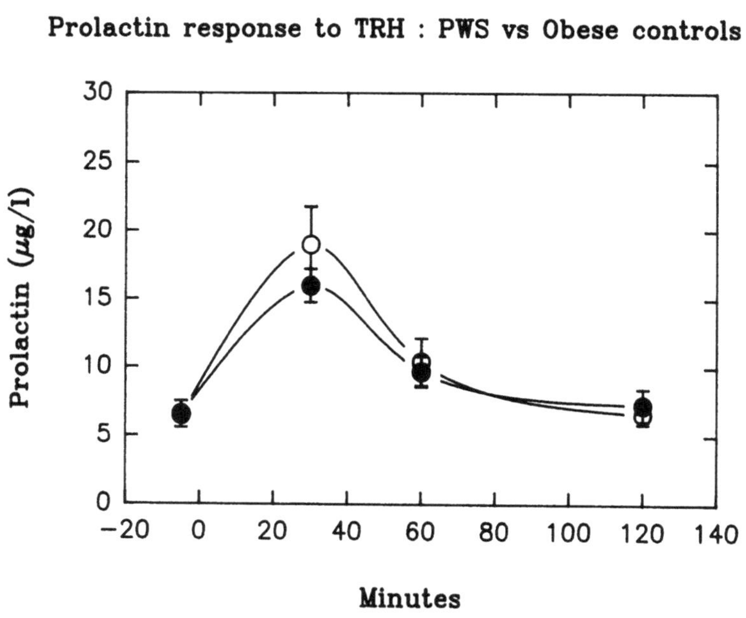

Figure 5. Prolactin levels in blood from PWS (o-o) and control obese (●-●) patients, following a bolus dose of TRH.

Glucose homeostasis

Baseline glucose and insulin levels in blood were the same in both groups. In order to test the ability to release insulin during prolonged hyperglycemia, glucose infusion was continued after the initial bolus dose. The intention was to "clamp" the blood glucose concentration to 18-20 mmol/l.

The obese children serving as controls all showed the expected rapid initial release of insulin, followed by a sustained and slow increase during the infusion. In all but one, insulin levels promptly decreased after one hour following glucose infusion (Figure 7). In the PWS group, the response was more variable (Figure 8). In one patient (H.E.), glucose levels were maintained at 14-18 mmol/l during the infusion but insulin never exceeded 30 μU/ml. Another patient (C.J.) reached the highest glucose concentrations in blood after the bolus dose (26 mmol/l) and during the subsequent one hour infusion (22-28 mmol/l). However, her insulin levels never exceeded 65 μU/ml. After the infusion, glucose levels fell at a normal rate, in spite of the comparative low insulin concentrations. In all of the PWS patients but only in one of the obese controls insulin levels failed to drop during the first 20 minutes after stopping the glucose infusion (Figure 9). Except for the first PWS patient (H.E.) discussed above, there was no difference between the rate of fall of blood glucose after the infusion between the groups.

DISCUSSION

The endocrine functions of the presently studied PWS patients confirmed previous notions of hypogonadotropic hypogonadism and short stature (Jeffcoate et al, 1980; Bray et al, 1983). It should be emphasized that PWS children may be of normal height before age 10, and still be very short in adulthood. This is reflected in the present study by the difference in mean SDS in those younger than 10 years (-1.4 SDS)

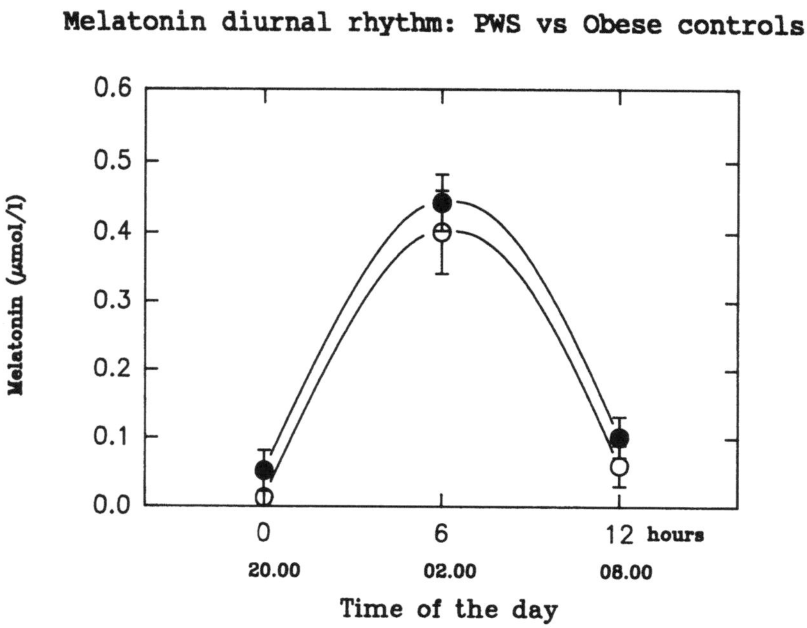

Figure 6. Melatonin levels in blood drawn from PWS (o-o) and control obese (●-●) patients at 20.00, 02.00 and 08.00 hrs.

and those older than 16 years (-3.8 SDS). Over the same ages, the body mass index increased from 25.7 to 43.3. The low growth hormone response to provocative tests (Bray et al, 1983) and the low urinary GH excretion (Blichfeldt et al, this Workshop) could be related to the obesity. In the present study, only the GH levels after falling asleep were significantly lower than those in the obese controls. However, only two out of seven PWS patients reached the level (10 μg/l) used as an arbitrary cut-off limit for "normal" response during AITT. It is well known that obese patients show low concentrations of GH in blood in spite of normal or even increased growth rate. The recent publication by Costeff et al (1990) showing that even PWS patients of normal weight have low levels of GH in blood is interesting. However, even in this case it can be argued that the ratio between fat and lean body mass is markedly increased even in normal-weight children with PWS, and might thus influence growth hormone levels. Even if this is the case, the published pilot studies of GH treatment of PWS patients show that in

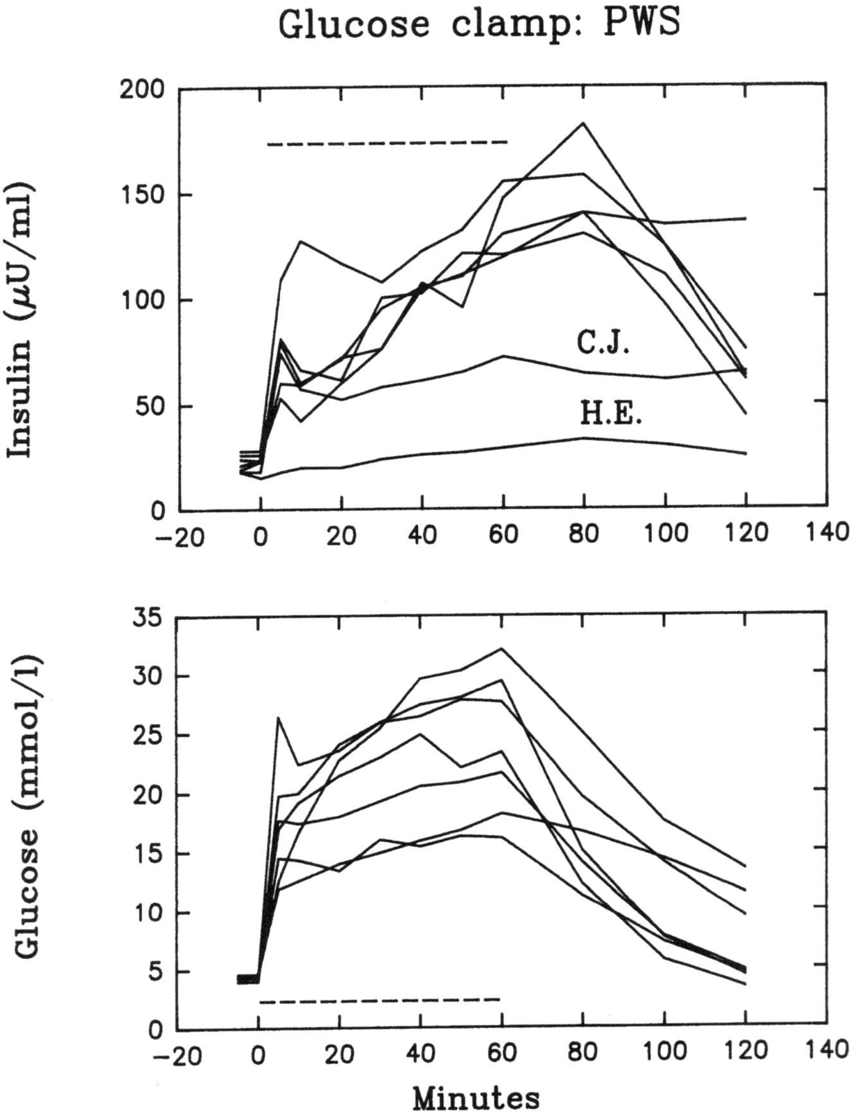

Figure 7. Individual insulin and glucose levels in PWS patients during a glucose "clamp". A bolus dose of glucose was given at 0 min, followed by a continuous infusion for 60 min (dotted line). Patients H.E. and C.J. are mentioned in the text.

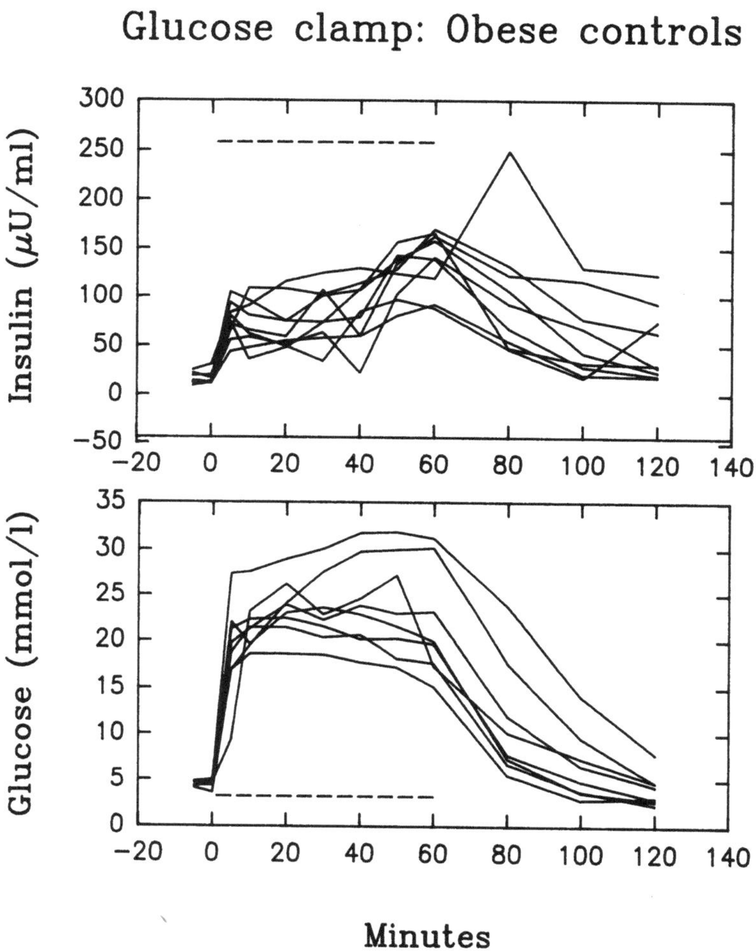

Figure 8. Individual insulin and glucose levels in control obese patients during a glucose "clamp". A bolus dose of glucose was given at 0 min, followed by a continuous infusion for 60 min (dotted line).

general growth rate is markedly increased during treatment (Lee et al, 1987; Pallante et al, 1990; Angulo et al, this Workshop).

The hypothalamic-pituitary-thyroid axis has previously been found to be normal in PWS. The present finding of a significantly increased secretion of TSH following TRH administration is interesting since this was also found in another group of patients with PWS (Jeffcoate et al, 1980). Peripheral levels of thyroid hormones were, however, not

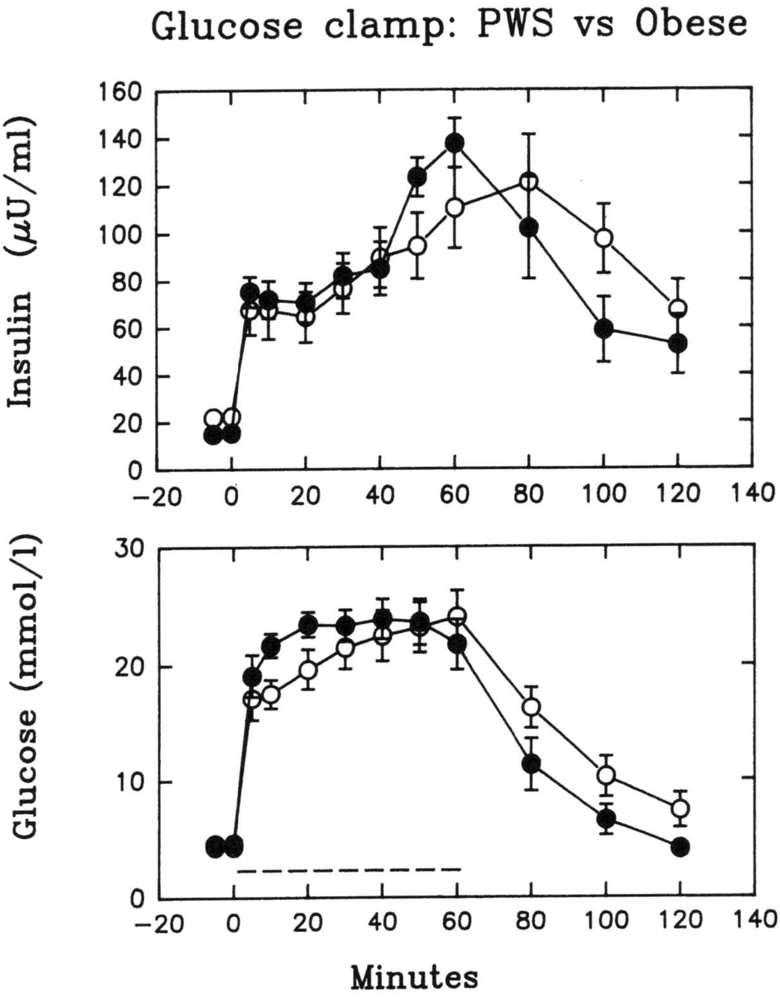

Figure 9. Insulin and glucose levels (means ± s.e.m.) in PWS (o-o) and control obese (●-●) patients during a glucose "clamp". A bolus dose of glucose was given at 0 min, followed by a continuous infusion for 60 min (dotted line).

significantly different.

Parents often report that their child with PWS has unusual sleeping habits, with a disturbed sleep rhythm. Although both the PWS and the control individuals in the present study had an unusual cortisol rhythm (Figure 3), there was no difference between the two groups in either melatonin or cortisol levels at night or in the morning. This confirms a previous report by Willig et al (1986).

The most interesting novel observation in the present study is that of insulin levels during and immediately after the glucose infusion test. There was no difference between the PWS and the obese controls in the initial rapid release of insulin following the first bolus injection of glucose, and there was no difference in the insulin levels in blood during the maintained high glucose concentrations. However, in the obese controls insulin levels rapidly declined following cessation of the glucose infusion. In the PWS individuals, insulin remained high for the first 30 minutes after glucose infusion in spite of a normal fall in glucose levels. This might indicate a slow turn-off of insulin secretion or, alternatively, a low clearance rate of circulating insulin. It should be investigated whether this prolonged hyperinsulinemia following a glucose load (or a meal?) is involved in the prolonged eating drive noted in PWS.

Endocrine therapy in PWS patients should be no different than that of other patients with endocrine insufficiencies. The frequently occurring cryptorchidism may at least in part be due to pre- and postnatal hypogonadotropic hypogonadism. Thus it seems logical to try hormonal treatment before surgery in PWS patients with cryptorchidism. This also has the advantage of avoiding anesthesia, which may be complicated in PWS.

Hypogonadism in puberty and early adulthood is variable. Most men with PWS show partial hypogonadism. Although sexual drive may not be a primary goal in PWS, androgen replacement may improve general activity, independence and determination. It should, however, be borne in mind that many adolescents with PWS have temper tantrums that might be aggravated by androgen treatment. In spite of this, it is our belief that the beneficial effects of androgen substitution are so great that it should be tested in adolescent and young adult men with PWS, provided that hypogonadism has been demonstrated. In young women with PWS, hypogonadism is a rule (Jeffocate et al, 1980; Bray et al, 1983; Cassidy, 1984). However, substitution therapy with estrogens (or progesterone) seems less important than androgen replacement to men; breast size is generally acceptable due to the obesity, and menstrual periods may not be desired. Furthermore, estrogens might increase obesity in some individuals. The new observation of frequent osteoporosis in PWS (Jones, this Workshop) might lead to second thoughts on estrogen replacement, although in obese patients peripheral production of estrogens should be sufficient to counteract osteoporosis in women without PWS.

Finally, it has been mentioned above that individuals with PWS may be deficient in growth hormone, and that GH administration in childhood results in improved growth rate. So far, there is little information on the effect of GH on metabolic parameters in PWS. Since individuals with PWS suffer not only from short final stature but also from osteoporosis and poor muscle development, further clinical studies should be encouraged. However, there is yet too little known about potential side effects to recommend GH administration to individuals with PWS outside clinical trials.

In summary, patients with PWS should be carefully followed to detect growth failure, hypogonadism, osteoporosis and impaired glucose homeostasis. When deficiences are found, substitution therapy should be considered.

REFERENCES

Angulo M, Castro-Magana M, Uy J, Rosenfeld W Growth hormone evaluation and treatment in Prader-Willi syndrome. This volume.

Blichfeldt S, Main K, Ritzén EM, Skakkebaek N-E Diminshed 24-hour urinary growth hormone excretion in patients with Prader-Willi syndrome. This volume.

Bray GA, Dahms WT, Swerdloff RS, Riser RH, Atkinson RL, Carrel RE (1983) The Prader-Willi syndrome: A study of 40 patients and a review of the literature. Medicine 62:59-80

Cassidy SB (1984) Prader-Willi syndrome. Current Problems in Pediatrics 14:1-63

Costeff H, Holm VA, Ruvalcaba R, Shaver J (1990) Growth hormone secretion in Prader-Willi syndrome. Acta Paediatr Scand 79:1059-1062

Grill V, Efendic S (1987) Studies on high and low insulin responders with the hyperglycemic clamp technique. Metabolism 36:1125-1131

Hall K, Enberg G, Ritzén M, Svan H, Fryklund L, Takano K (1980) Somatomedin A levels in serum from healthy children and from children with growth hormone deficiency or delayed puberty. Acta Endocrinol 94:155-165

Jeffocate WF, Laurance BM, Edwards CRW, Besser GM (1980) Endocrine function in the Prader-Willi syndrome. Clin Endocrinol 12:81-89

Jones MW Scoliosis surgery in the Prader-Willi syndrome. This volume.

Lee PDK, Wilson DM, Hintz RL, Rosenfeld RG (1987) Growth hormone treatment of short stature in Prader-Willi syndrome. J Pediatr Endocrinol 2:31-34

Pallante VA, Lovinger RD, Bodurtha J (1990) Growth hormone treatment of short stature in Prader-Willi syndrome. Abstract presented at 41st Annual Meeting of the American Society of Human Genetics, Cincinnati, October 1990. Am J Hum Genet Suppl 47:A69

Povoa G, Roovete A, Hall K (1984) Cross reaction of serum somatomedin binding protein in a radioimmunoassay developed for somatomedin binding protein isolated from human amniotic fluid. Acta Endocrinol 107:563-570

Prader A, Labhart A, Willi H (1956) Ein Syndrom von Adipositas, Kleinwuchs, Kryptorchismus, und Oligophrenie nach myatonierartigen Zustand im Neugeborenalter. Schweiz Med Wochenschr 86:1260-1261

Willig RP, Braun W, Commentz JC, Stahnke N (1986) Circadian fluctuation of plasma melatonin in Prader-Willi's syndrome and obesity. Acta Endocrinol Suppl 279:441-415

ACKNOWLEDGMENTS

These studies were supported by the Swedish Medical Research Council and the "Expressen" Prenatal Research Fund. Professor Lennart Wetterberg is gratefully acknowledged for the assay of melatonin.

GROWTH HORMONE EVALUATION AND TREATMENT IN PRADER-WILLI SYNDROME

Moris Angulo
Winthrop University Hospital
State University of New York, Stony Brook
Department of Pediatrics
Division of Endocrinology & Genetics
Mineola, New York
USA

Mariano Castro-Magana, John Uy, Warren Rosenfeld.

Winthrop University Hospital, Mineola, New York

INTRODUCTION

Prader-Willi syndrome (PWS) is characterized by infantile hypotonia, obesity, hypogonadism, mental deficiency, short stature, characteristic facial features, small hands and feet, and an interstitial deletion of chromosome 15(q11-13) in about one-half of the patients. Short stature and delayed skeletal maturation occur in 90% of children with PWS. The basis for the short stature, however, remains uncertain. Pituitary evaluation of growth hormone (GH) deficiency by pharmacological tests have varied from normal to impaired in PWS children (Bray et al, 1983; Lee et al, 1982).

METHODS

We evaluated the spontaneous 24-hour growth hormone (GH) secretion and the GH response to pharmacological stimuli with clonidine, L-dopa and insulin-induced hypoglycemia in 15 obese and four non-obese children, aged 1 to 15.5 years (Table). Ten of these children have received recombinant human GH, 0.1 mg/kg body weight three times a week intramuscularly, for at least six months.

NATO ASI Series, Vol. H 61
Prader-Willi Syndrome
Edited by S. B. Cassidy

RESULTS

Interval growth increased from 2.0 ± 0.20 cm to 5.3 ± 1.5 cm after six months therapy without advancing bone age more than chronological age. Weight gain during the same period of time decreased from 3.6 ± 0.9 kg to 1.3 ± 0.7 kg, whereas somatomedin-C (SM-C) increased significantly from 0.6 ± 0.3 U/ml to 1.5 ± 0.2 U/ml (Figure).

SEX	CA (yr)	BA	WT. (kg)	HT. (cm)	15q-	SM-C (U/ml)	GH peak(ng/ml) I	C	D	I-24 (hr)
M	1.0	0.8	6.6	71	+	0.2	10.0	2.5	3.0	1.9
F*	3.5	2.5	14.8	87	+	0.4	10.6	3.9	2.7	1.8
M*	5.0	3.5	21.0	104	+	0.7	19.3	6.6	3.7	2.1
M*	8.0	6.5	40.0	124	-	0.5	2.6	1.6	4.9	1.0
M	12.5	11.0	47.2	132	-	1.1	13.9	3.1	12.6	1.0
M*	10.3	9.0	56.7	145	+	0.4	19.6	2.1	1.5	1.5
F*	7.0	5.5	35.5	109	+	1.0	20.4	3.2	11.4	1.5
M	14.0	14.0	90.5	156	+	1.0	3.6	1.5	7.5	1.0
M*	15.5	14.5	84.5	151	-	1.1	4.	12.7	6.0	1.9
M*	10.0	9.0	62.8	128	+	0.5	2.5	2.3	1.6	2.1
M	7.0	5.5	51.5	114	+	0.4	6.0	2.0	2.0	1.7
M*	8.5	7.5	36.0	121	-	1.2	8.2	7.2	2.3	1.9
M*	4.5	3.7	17.0	96	+	0.3	6.0	4.2	11.4	1.5
M*	12.5	12.5	66.0	144	-	0.2	6.4	2.4	2.1	1.7
F	7.0	5.7	30.5	110	+	0.4	8.0	2.7	5.6	1.5
M	10.0	7.0	25.3	110	+	0.6	2.5	1.5	2.9	1.5
M	3.0	2.5	15.0	88	+	0.7	14.7	3.1	11.4	2.2
M	6.0	5.0	31.0	112	+	0.5	1.5	1.5	1.5	1.5
F	3.7	2.7	21.3	97	+	1.0	3.2	5.1	1.5	1.5

* Patients placed on GH therapy after testing. 1-24h: Integrated 24-hr GH secretion
CA: Chronological age; BA: Bone Age; I: Insulin ; C: Clonidine ; D: L-dopa

Table I. PWS patients clinical data and GH reponse to pharmacological and physiological tests

In spite of the spectrum of GH response to the three pharmacological tests, the integrated 24-hour GH secretion was abnormally low (<3.5 ng/ml) in all of them. These findings suggest hypothalamic neurosecretory irregularities of GH release. The improvement in growth rate after GH therapy was similar to that seen in children with classical GH deficiency.

Significant decrease in weight gain associated with loss of adiposity was also observed after GH treatment. Other anabolic actions of GH such as increase in lean mass, muscle strength, exercise capacity, and energy expenditure in GH deficient individuals have been reported by other investigators (Prader et al, 1972; Crist et al, 1987; Henneman et al, 1960; Jorgensen et al, 1989).

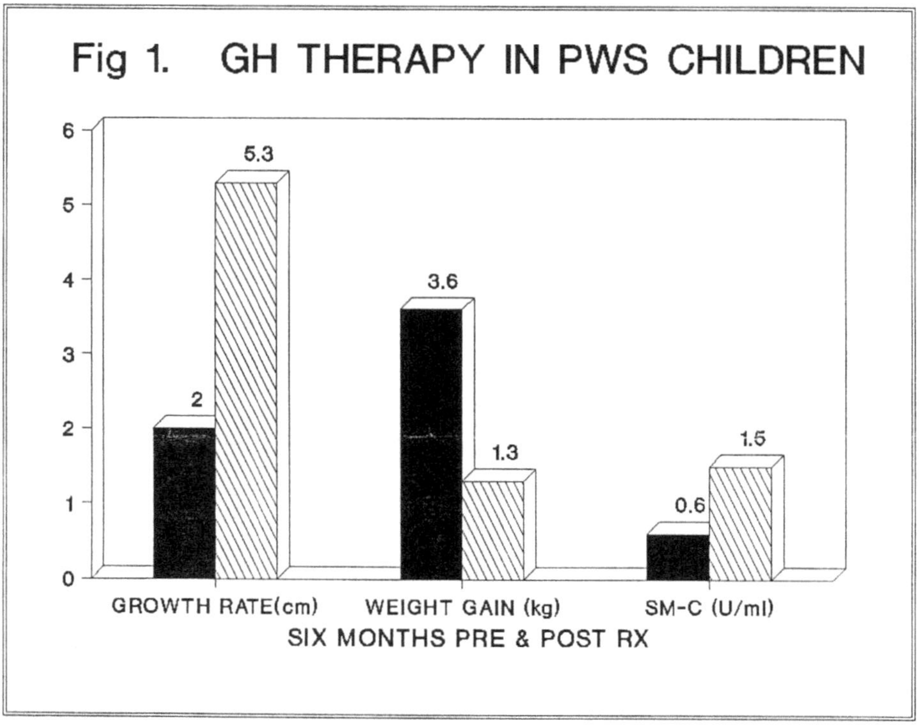

DISCUSSION

Neurosecretory GH deficiency could be considered to contribute, at least in part, to the short stature of PWS, and may add further evidence of hypothalamic dysfunction characteristic of individuals with this disorder. The improvement in growth velocity and decrease in weight gain which were seen in our patients, in addition to other known anabolic effects of GH, make children with PWS potential candidates for GH therapy.

REFERENCES

Bray GA, Dahms WT, Swerdloff RS, Fiser RH, Atkinson RL, Carrell RE (1983) The Prader-Willi syndrome: A study of 40 patients and review of the literature. Medicine 62:59-80

Lee RDK, Wilson DM, Hintz RL, Rosenfeld RG (1987) Growth hormone treatment of short stature in Prader-Willi syndrome. J Ped End 2:31-34

Prader A, Ferrandez A, Zachmann M, Illig R (1972) Effect of hCG treatment on growth, bone age and skin fold tickness in 44 children with growth hormone deficiency. In: Pecile A, Muller EE (eds). Growth and Growth Hormone. Amsterdam: Excerpta Medica, p 457

Crist DM, Peake GT, Mackinnon LT, Kraner JC (1987) Exogenous growth hormone (GH) treatment alters body composition and increases natural killer cell activity in women with impaired endogenous GH secretion. Metabolism 36:1115-1117

Henneman PH, Forbes AP, Moldawer M, Dempsey EF, Carroll EL (1960) Effects of human growth hormone in man. J Clin Invest 39:1223-1238

Jorgensen JOL, Pedersen SA, Tuesen L, Jorgensen J. Ingermann-Hansen T, Skakkebaek NE, Christiansen JS (1989) Beneficial effects of growth hormone in treatment in GH-deficient adults. Lancet 1:1221-1224

DIMINISHED 24 HOUR URINARY GROWTH HORMONE EXCRETION IN PATIENTS WITH PRADER-WILLI SYNDROME

Susanne Blichfeldt, M.D.
Department of Growth and Reproduction
Rigshopitalet
2100 Copenhagen
Denmark

K. Main[1], M. Ritzén[2], N.E. Skakkebaek[1].

Department of Growth and Reproduction, Rigshopitalet, Copenhagen, Denmark[1]; Paediatric Endocrine Unit, Karolinska Institute, Stockholm, Sweden[2].

INTRODUCTION

Short stature in Prader-Willi syndrome (PWS) has been suggested to be due to growth hormone deficiency (GHD) (Bray, 1983). GHD has been documented in a small number of patients with PWS (Costeff et al, 1990), and a few children have been reported to respond well to growth hormone (hGH) treatment (Lee et al, 1987). Urinary growth hormone (U-GH) excretion reflects to some extent the endogeneous growth hormone (GH) production (Girard et al, 1990). Collection of urine samples is easy, can be done at home and is confortable for the patient.

The aim of this study was to investigate whether u-GH excretion is altered in PWS.

STUDY POPULATION

The study group comprised 21 patients with PWS; 12 children, aged 4-18 years (4 male, 8 female) and 9 adults, aged 19-31 years (5 male, 4 female). Growth curves for PWS (Holm, 1988) showed that the patients all had height within the expected range for PWS (Figure 1).

NATO ASI Series, Vol. H 61
Prader-Willi Syndrome
Edited by S. B. Cassidy

Height and weight:height ratios of the patients were compared to normal percentiles in the Danish population (Andersen et al, 1982) (Figure 2). Ten of 12 children (83%) had a height below the 3rd centile of normal values. The weight:height ratio varied from the 10th to above the 97th percentile, with the median being the 94th percentile.

The lowest weight:height ratios were found among the youngest children, due to early dietary restrictions. Eight of 9 adults had heights below the 3rd centile, only one being within the normal range. Two adults were on substitution therapy with sex steroids.

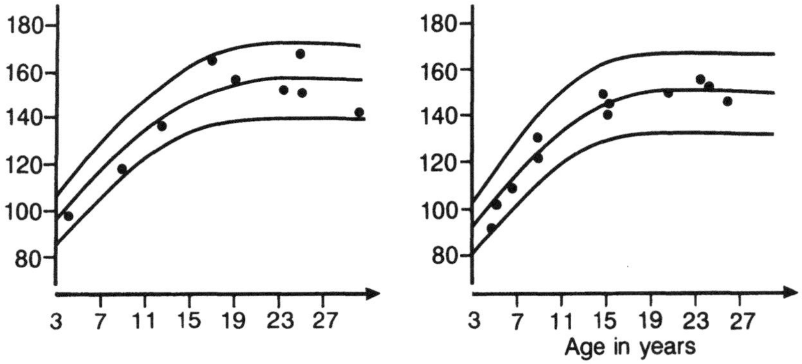

Figure 1: Heights (cm) of patients in the study group plotted onto curves for PWS; upper curve: males, lower curve: females.

METHODS

From each participant a 24 hour urine was collected. Growth hormone in urine was analyzed by ELISA technique (Main et al, 1991). The U-GH vales (ng/24h) of the patients were compared to a control group of 584 healthy Danish children and 110 healthy adults (Main et al, 1991).

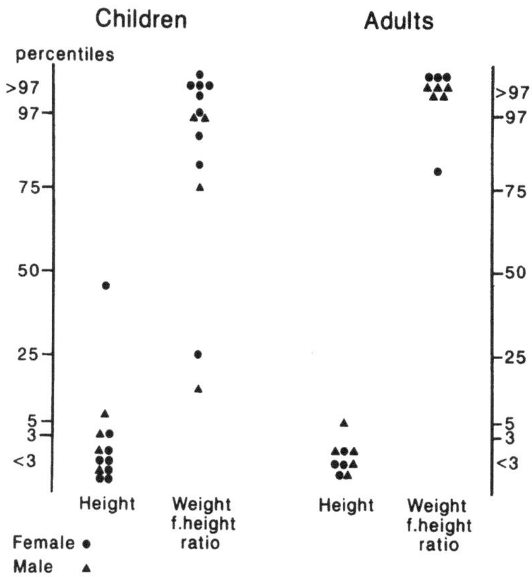

Figure 2: Heights and weight:height ratios in the patients compared to normal values for Danish population.

RESULTS

The results of the study are shown in Figures 3a and 3b. In the children's group, all but one had a U-GH excretion below the 50th percentile. Eight of 12 had values below the 25th percentile, and 4 had values below the 5th percentile. The youngest children tended to show the lowest U-GH excretion values. In the adult group all but one had U-GH values below the 50th percentile, but none of the adults had values below the 3rd percentile.

DISCUSSION

Urinary GH excretion in our group of 21 patients with PWS

tended to show low to subnormal values in children and low values in adults. The fact that subnormal values were only found in the children's group might be due to the influence of sex steroids on U-GH excretion in adulthood.

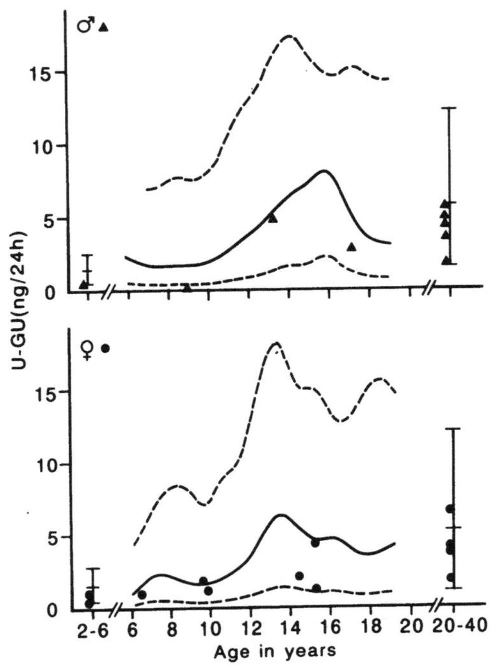

Figure 3a: 24 hour U-GH excretion in patients with PWS compared to the excretion in healthy Danish children and adults. Curves indicate 3rd, 50th and 97th percentiles.

It has been suggested that subnormal GH values in PWS might be due to obesity, since it is well known that serum GH values in obese patients of normal height are low both in profiles and during provocation tests (Bierich, 1987). In this study, the reference population was of normal weight for height and we did not include a control group of obese but otherwise healthy participants. However, it has to be emphasized that in our patients the lowest U-GH values were found in children with normal weight, but retarded growth.

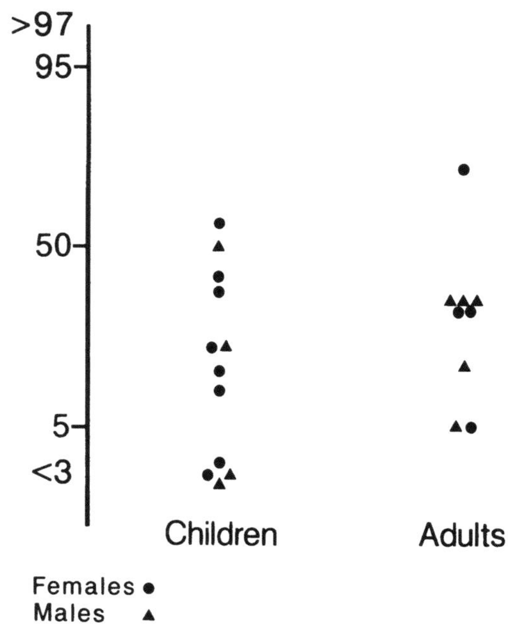

Figure 3b: Percentiles of U-GH excretion (ng/24h) in patients with PWS as compared to healthy individuals.

CONCLUSION

This study lends further support to the hypothesis that PWS may be associated with some degree of GH insufficiency. It remains to be seen to what extent U-GH excretion is influenced by sex steroid and/or obesity in PWS.

REFERENCES

Andersen E, Hutchings B, Jansen J, Nyholm M (1982) Højde og vaegt hos danske børn. Ugeskr Laeger 114:1760-1765

Bierich JR (1987) Serum growth hormone levels in provocation tests and during nocturnal spontaneous secretion: A comparative study. Acta Paed Scand (Suppl) 337:48-59

Bray GA (1983) The Prader-Willi syndrome: A study of 40 patients and a review of the literature. Medicine 62(2):59-80

Costeff HM, Holm VA, Ruvalcaba R, Shaver J (1990) Growth hormone secretion in Prader-Willi syndrome. Acta Paed Scand 79:1059-1062

Girard J et al (1990) Urinary measurement of growth hormone secretion. Acta Paed Scand (Suppl) 336:149-154

Holm VA (1988) Growth charts for Prader-Willi syndrome. In: Greenswag LR, Alexander RD (eds) Management of Prader-Willi syndrome. Springer-Verlag, New York, p 237-240

Lee PDK, Wilson DM, Rountree L, Hintz RL, Rosenfeld RG (1987) Linear growth response to exogeneous growth hormone in Prader-Willi syndrome. Am J Med Genet 28:865-871

Main K, Philips M, Jørgensen M, Skakkeaek NE (1991) Urinary growth hormone (U-GH) excretion in 694 healthy children and adults. Unpublished data

ENERGY EXPENDITURE IN THE PRADER-WILLI SYNDROME

Peter S.W. Davies
Infant and Child Nutrition Group
MRC Dunn Nutrition Unit
Downhams Lane
Milton Road
Cambridge, England

C. Joughin[1], M.B.E. Livingstone[2], and N.D. Barnes[3]

Infant and Child Nutrition Group, MRC Dunn Nutrition Unit, Cambridge, England[1]; Biomedical Sciences Research Centre, University of Ulster, Northern Ireland, United Kingdom[2]; Department of Paediatrics, Addenbrookes Hospital, Cambridge, England[3]

INTRODUCTION

The Prader-Willi syndrome is characterised by hypotonia in the neonatal period, hypogonadism, short stature, small hands and feet, mental retardation and obesity. However, it is unclear whether the obesity is due primarily to an excessive energy intake, a reduced total energy expenditure or a combination of both these factors (Bray et al, 1983; Coplin et al, 1985). Studies that might contribute knowledge to the question have been hampered by the inability to measure some of the parameters, notably total energy expenditure, especially in children.

Total energy expenditure can now be measured non-invasively in "free-living" infants, children and adults using the doubly labelled water technique. This method thus allows studies of energy metabolism in groups of individuals in whom such measures have been previously impossible.

Therefore, we have measured total energy expenditure and resting metabolic rate in a cohort of children with the Prader-Willi syndrome (PWS).

NATO ASI Series, Vol. H 61
Prader-Willi Syndrome
Edited by S. B. Cassidy

METHODS

Nineteen subjects with PWS were recruited with the assistance of the Prader-Willi Syndrome Association (UK). All the subjects were living at home at the time of the study and eating self- selected or parentally selected diet.

Total energy expenditure (TEE) was measured using the doubly labelled water technique (Coward and Prentice, 1985). Two stable isotopes of water ($H_2{}^{18}O$ and ^2H_2O) were administered orally. In all cases the dose was most easily administered via a standard drinking straw from a 125 ml bottle.

In each case the weight of the dose given was determined accurately. A dose consisting of 0.125 g/kg body weight of $H_2{}^{18}O$ and 0.1 g/kg body weight of ^2H_2O was used. A single urine sample was collected before administration of the dose to determine natural concentrations of ^2H_2O and $H_2{}^{18}O$ in body fluids. Urine samples were then collected 5 hours post-dose and thereafter every 24 hours for 10 or 14 days depending on the age of the child. The enrichment of the two isotopes was measured in the urine samples using isotope ratio mass spectrometry. Briefly, the principle of the method, which is described in detail elsewhere (Coward and Prentice, 1985) is as follows. Labelled hydrogen (^2H) is excreted from the body in water and hence its disappearance from body fluids reflects water production rate. Labelled oxygen (^{18}O) will equilibrate between water and carbon dioxide because of the action of carbonic anhydrase, and so will be excreted both in water and expired CO_2. Thus the difference between the disappearance rates of the two isotopes is proportional to CO_2 production rate. Assuming a respiratory quotient of 0.85 and a fractionation factor of 0.25 in all individuals, TEE can be calculated using Weir's formula (Weir, 1949). Further details of the calculations are described by Davies et al, (1989). Body weight was recorded to the nearest 50 g on electronic scales.

Resting metabolic rate (RMR) was measured using a portable open circuit indirect calorimeter (Crisp et al, 1991). Subjects were required to rest for approximately 20 minutes before the measurement. Room air was drawn through a clear, plastic hood which was placed over the subject's head. Samples of gas leaving the hood were collected after 10 to 15 minutes. Ambient gas samples were collected simultaneously. Analysis of the carbon dioxide and oxygen concentration in the expired and ambient samples and flow rate of the air through the hood allows the calculation of energy expenditure.

Basal metabolic rate (BMR) was also predicted from height and weight using Schofield's equation (Schofield, 1985).

STATISTICAL ANALYSIS

Studies involving measurements of TEE in disease states have often lacked good control data. Such a data set may now exist (Davies, unpublished results; Davies et al, 1989; Davies et al, 1990). These data consist of 155 measurements of TEE in children aged 3-18 years. The first data set was produced by longitudinal measurements of TEE in a cohort of 50 normal, healthy babies recruited from The Rosie Maternity Hospital in Cambridge, United Kingdom. The second data set was cross-sectional and consisted of children recruited from nurseries, schools and college in Cambridge and in Belfast, United Kingdom.

In comparative work it is important that TEE be expressed correctly and that the influence of confounding factors such as age, sex, body weight and body composition are negated.

Appropriate statistical analysis showed that in the normal control data the effect of age, sex and body composition was removed if TEE was expressed as (kcal/kg fat free mass $^{0.75}$). Fat free mass (FFM) was measured via knowledge of the $H_2{}^{18}O$ dilution space calculated during the doubly labelled water technique. Reference values (Fomon et al, 1982) were used for the level of hydration of the FFM. Body fat mass was calculated as the difference between body weight and FFM.

RESULTS

Some of the physical characteristics of the 19 children with PWS are shown in Table I. The mean and SD of TEE expressed as kcal/kg body weight/day and kcal/kg $FFM^{0.75}$/day are shown in Table II.

Table I. Physical Characteristics of the Studied Children with PWS

	Mean	SD
Age (y)	8.8	4.4
Weight (kg)	34.3	26.6
Height (cm)	120.9	23.6
% Body Fat	43.0	10.3

Table II. Total Energy Expenditure in the Studied Children with PWS

	Mean	SD
Total Energy Expenditure (kcal/day)	1439	589
Total Energy Expenditure (kcal/kg/day)	47.7	15.4
Total Energy Expenditure (kcal/kg $FFM^{0.75}$/day)	165.2	24.9

Regression analysis showed that when expressed as kcal/$FFM^{0.75}$/day, the mean energy expenditure of the children with PWS was 13% below the mean for the control data set. This was a highly significant deficit (t= -3.6, p<0.01).

The mean RMR in kcal/day and the mean predicted BMR for all the recruited children are shown in Table III. The mean and standard deviation of the TEE:predicted BMR is also shown in Table III. Data were not available for the entire group as 8 of the children were either too apprehensive or too restless for a valid RMR to be measured. There was no significant difference between measured RMR and predicted BMR.

Table III. Measured Resting Metabolic Rate, Predicted Basal Metabolic Rate, and the Ratio Between Total Energy Expenditure and Predicted BMR in the Tested Patients with PWS.

	Mean	SD
Measured RMR (kcal/day)	1302	405
Predicted BMR (kcal/day)	1323	445
TEE:Predicted BMR	1.36	0.20

DISCUSSION

Obesity in childhood and later life is a frequent finding in the Prader-Willi syndrome. There have been conflicting reports as to whether the accumulation of body fat is due entirely to the hyperphagia that is characteristic of the syndrome or whether energy balance is further compromised by a reduction in energy expenditure. In this study, mean TEE was significantly reduced in 19 children and adolescents with PWS, in comparison with a large cohort of normal children, after taking into account body composition, sex and age.

Moreover, after adjusting for age, fat free mass and sex by multiple regression analysis, the resting metabolic rates of the children with PWS were not statistically different from the control children (t=0.19).

It could be argued that the predicted BMR might be greater than measured BMR or RMR in the children with PWS. For any given body weight, the children with PWS tend to have a greater percentage body fat than the children used to create the prediction equation. Thus, these children will have smaller amounts of metabolically active lean tissue for any given body weight, and thus a lower BMR. Also, BMR reduces with age, and the larger body weights of the children with PWS would normally be associated with older children with lower BMR. The mean TEE:BMR ratio in children with PWS was 1:36. This is very low in comparison with normal children and adolescents (Davies et al, 1990). These low data are a consequence of the reduced TEE in conjunction with the essentially normal BMR.

Equally low values for the TEE:BMR ratio have been reported before in adults (Prentice et al, 1985). Nevertheless, using a factorial method for the calculation of TEE by summing the energy costs of various activities as multiples of BMR (FAO/WHO/UNU, 1985), a TEE only 26% greater than BMR is difficult to achieve using existing data. Allowing for 5 hours each day at school, at energy cost of BMR x 1.55 (FAO/WHO/UNU, 1985), the remaining 19 hours would need to be spent very close to BMR if total ratio of 1.36 is to be achieved. It is possible that children with PWS are significantly less active within school or that the periods of activity that do occur are of limited duration and intensity.

If activity levels could be increased in children with PWS, this may be an important contribution to the control of body weight and body composition.

REFERENCES

Bray GA, Dahms WT, Swerdloff RS, et al (1983) The Prader-Willi syndrome: A study of 40 patients and a review of the literature. Medicine 62:59-80

Coplin SS, Hine J, Gormican A (1976) Outpatient dietary management in the Prader-Willi syndrome. J Am Diet Assoc 68:330-334

Coward WA, Prentice AM (1985) Isotope method for the measurement of carbon dioxide production rate in man. Am J Clin Nutr 41:659-663

Crisp JA, Murgatroyd PR, Venman M, Davies PSW (1991) Field measurement of resting metabolic rate in neonates. Submitted: Brit J Nutr

Davies PSW, Ewing G, Lucas A (1989) Energy expenditure in early infancy. Brit J Nutr 62:621-629

Davies PSW, Cole TJ, Lucas A (1989) Adjusting energy expenditure for body weight in early infancy. Euro J Clin Nutr 43:641-645

Davies PSW, Livingston MBE, Prentice AM, Coward WA, Jagger SE, Stewart C, Whitehead RG (1990) Total energy expenditure during childhood and adolescence. Proc Nutr Soc

FAO/WHO/UNU (1985) Energy and protein requirements. WHO technical report series, No. 724

Fomon SJ, Haschke F, Ziegler EE, Nelson SE (1982) Body composition of reference children from birth to age 10 years. Am J Clin Nutr 35:1169-1175

Prentice AM, Coward WA, Davies HL, Murgatroyd PR, Black AE, Goldberg GR, Ashford J. Sawyer M, Whitehead RG (1985) Unexpectedly low levels of energy expenditure in healthy women. Lancet II:1419-1422

Schofield WN (1985) Predicting basal metabolic rate, new standards and review of previous work. Hum Nutr Clinc Nutr 39C:Suppl 1:5-41

Weir JB (1949) New methods for calculating metabolic rate with special reference to protein metabolism. J Physiol 109:1-9

ANTERO-POSTERIOR CEPHALOMETRIC ANALYSIS OF THE CRANIOFACIAL COMPLEX IN THE PRADER-WILLI SYNDROME

Andrew E. Poole
Department of Pediatric Dentistry
Craniofacial Disorders Team
University of Connecticut Health Center
Farmington, Connecticut
USA

Rima M. Munoz[1] and Suzanne B. Cassidy[2]

Department of Pediatric Dentistry, University of Connecticut Health Center, Farmington, Connecticut[1]; Department of Pediatrics, Section of Genetics/Dysmorphology, University of Arizona College of Medicine, Tucson, Arizona[2] USA

INTRODUCTION

Prader-Willi syndrome (PWS) is a multisystem disorder with characteristic clinical features including hypotonia, hypogonadism, insatiable appetite, truncal obesity, mild mental deficiency, short stature, small hands and feet, and facial dysmorphology. The typical facies are described as a combination of the following features: almond-shaped palpebral fissures which may be upslanting, long narrow face, downturned mouth, and thin upper lip (Cassidy, 1984).

Hall and Smith (1972) were the first to suggest that narrow bifrontal diameter, defined as the distance between the external rim of the orbit on the frontal bones, was a major feature of this syndrome, particularly in younger patients. Butler and Meaney (1987) supported previous suggestions that dolichocephaly, defined as elongation in the shape of the skull such that the cephalic index (CI) is 75.9 or less, may be an early diagnostic feature of PWS. They also confirmed that the bifrontal diameter was narrow. Schaedel et al (1990), using lateral cephalometric analysis, stated that the small CI must be the result of a narrow, rather than a long, face since the facial height was within normal limits. The purpose of the present study was to

NATO ASI Series, Vol. H 61
Prader-Willi Syndrome
Edited by S. B. Cassidy

analyze antero-posterior radiographs of patients with PWS to confirm the observation of Schaedel et al (1990) and to determine the usefulness of this technique in quantitatively diagnosing this disorder.

MATERIALS AND METHODS

Population

The study population consisted of 15 patients, 6 males and 9 females, previously diagnosed with PWS by one of us, a clinical geneticist (SBC), using clinical examination and chromosome studies. The characteristics of this population are shown in Table I.

Table I. Description of the Population and Radiographic Analysis Used in This Study

Population

Number:	15 patients
Cytogenetic findings:	13 with 15q deletion; 2 normal
Sex:	6 males, 9 females
Age:	6 children (4 years to 13 1/12 years) 9 adults (18 8/12 years to 49 1/12 years)

Radiographic Method

The standard technique for measurement of bony interorbital distance (BIOD) from antero-posterior radiographs was devised by Currarino and Silverman in 1960, and modified by Morin et al (1963). The method of Costavas et al (1982), with revision by Aduss et al (1987), were used for all other measurements. In order to minimize the rotation of the patient's head and to allow accurate superimposition, the x-rays were taken with the patient placed in a cephalometric headholder with the distance from the x-ray source kept constant.

Radiographic Analysis

The tracing of the typical antero-posterior cephalometric radiograph is shown in Figure 1, with the measurement points marked. These are identified as:

1) Euryon (Eu) - the most lateral point on the outline of the cranial vault;
2) Frontotemporal (FrTe) - the most lateral point on the frontal bone in the orbit;
3) Roof of the orbit (RO) - upper most point in the roof of the orbit;
4) Medial orbital margin (MO) - the most medial point on the bony outline of the orbit;
5) Interior orbital margin (IO) - the most inferior point on the outline of the orbit;
6) Zygoma (Zy) - the most lateral point on the contour of the mastoid process;
7) Gonion (Go) - the most lateral point on the mandibular angle;
8) Point A - point where the MO-MO plane intersects the RO-IO plane;
9) Menton (Men) - the lower most point in the contour of the chin.

The measurements computed from these points are listed in Table II.

Table II. Radiographic Analysis and Cephalic Index

Radiographic Analysis

Measurement	Location (Fig. 1)
Bony Interorbital Distance (BIOD)	Mo - Mo
Cranial Width	Eu - Eu
Bifrontal Diameter	FrTe - FrTe
LOCP	A - Ro
Bizygomatic Diameter	Zy - Zy
Bigonial Diameter	Go - Go

Calculation

$$\text{Cephalic Index (CI)} = \frac{\text{Head Breadth}}{\text{Head Length}} \times 100$$

Dolichocephaly = CI <75.9%

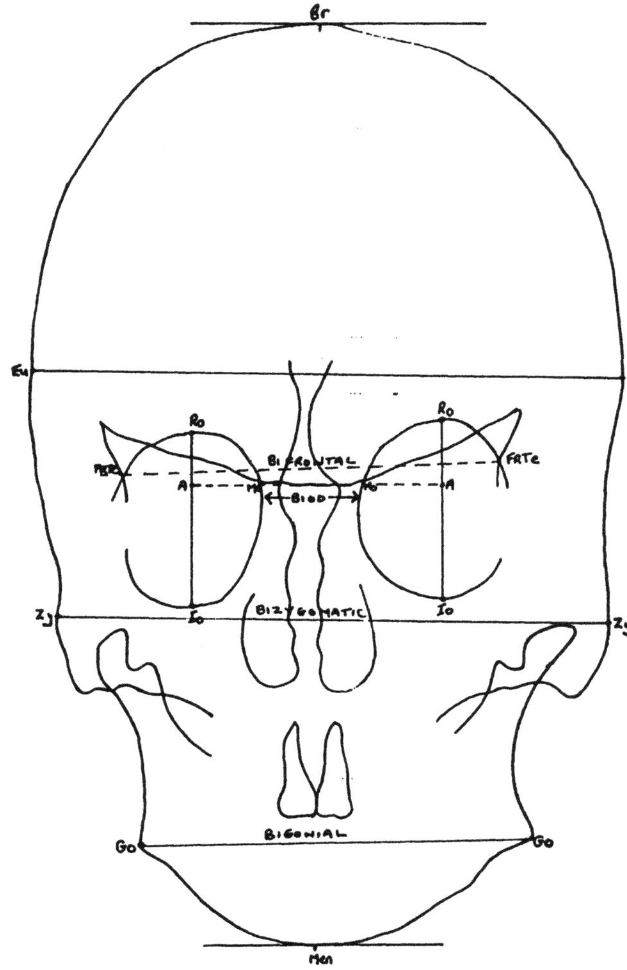

Figure 1. Tracing of antero-posterior cephalometric radiograph of a patient with PWS showing the points from which linear measurements used in this study were computed. (See text for definition of points.)

Standards for these measurements were obtained from the published sources listed above and from Farkas (1981) and Woods (1950). Only those measurements for which adequate normal age and sex matched controls were available for comparison were used in this study.

Statistical Analysis

Mean and standard deviation for the subjects were computed and compared with published mean values for all normal subjects. Paired t tests of differences between normal and age- and sex-matched control values were also computed and p values obtained.

RESULTS

Table III. Cephalic Index, Cranial Width and Bifontal Diameter in 15 Patients with PWS

Patient	Cephalic Index	Cranial Width(mm) Exp	Control	BifrontalDiameter(mm) Exp	Control
1	76	145	136	88	104
2	66	131	132	88	103
3	70	132	132	90	103
4	59	126	135	87	107
5	65	142	141	97	113
6	54	159	141	92	113
7	60	131	137	93	118
8	65	150	143	100	117
9	*	149	138	97	114
10	*	148	142	100	118
11	74	143	138	99	114
12	*	140	142	89	118
13	*	145	138	90	114
14	73	161	138	100	114
15	71	149	138	102	114
Mean		143.4	138.1	94.1	112.3
S.D.		10.1	3.4	5.3	5.4
Paired t value	:	-2.407		: 14.605	
p		<0.025		: 0.0005	

Table III shows CI, maximum cranial width and bifrontal diameter in the patients with PWS compared to normal controls. Ten of the 11 patients in which CI was measured had values of <75.9% indicating the presence of dolichocephaly. Patient #1 had a borderline value for dolichocephaly.

Cranial width had a mean value of 143.4 in patients with PWS compared to a mean value of 138.1 in the control subjects. The standard deviation in the patients with PWS was much larger (10.1) than in the normals (3.4), indicating a greater degree of variability.

In all PWS cases the bifrontal diameter was statistically narrow compared to the normal population (p = 0.0005) indicating that the subjects with PWS have very narrow foreheads.

The bony interorbital distance was increased in 13 of the 15 patients with PWS studied (Table IV). The mean BIOD value for all patients with PWS was 23.6 compared to 21.1 in the normal controls, and the values were statistically different at the p <0.005 level. Alternatively, the LOCP values were consistently larger in all patients with PWS, with a mean value of 44.7 compared to 30.0 for normals, and this was statistically different at the p = 0.0005 level.

Table IV. Bony Interorbital Distance (BIOD) and Level of Cribriform Plate (LOCP) in 15 Patient With PWS

	BIOD(mm)		LOCP (%)	
Patient	Exp	Control	Exp	Control
1	20	17	56	33
2	20	17	53	35
3	24	17	46	35
4	20	20	41	33
5	19	22	37	28
6	24	22	38	28
7	22	22	41	30
8	26	24	50	28
9	25	22	49	30
10	26	23	49	27
11	24	22	29	29
12	26	23	43	27
13	24	22	43	29
14	25	22	39	29
15	29	22	56	29
Mean	23.6	21.1	44.7	30
S.D.	2.8	2.3	7.6	2.7
Paired t value :	-3.861		: -7.903	
p	<0.005		<0.0005	

Table V shows the mean bizygomatic width for patients with PWS was 132.9 compared to 117.8 for normal subjects. The difference was statistically significant at the p = 0.0005 level indicating that this part of the face is wide in PWS. The bigonial diameter, which corresponds to the width of the mandible, showed more variability, but was large in 10 of the 15 subjects. The values, however, were generally in the range of those for the control population and were only statistically different at the p = 0.05 level.

Table V. Bizygomatic and Bigonial Diameter in 15 Patients with PWS

Patient	Bizygomatic(mm) Exp	Bizygomatic(mm) Control	Bigonial(mm) Exp	Bigonial(mm) Control
1	127	105	78	73
2	115	104	70	74
3	122	104	77	74
4	125	113	81	82
5	127	120	88	89
6	143	120	103	89
7	130	121	79	88
8	145	125	94	93
9	135	121	99	88
10	144	125	98	93
11	135	121	95	88
12	135	125	102	93
13	124	121	81	88
14	147	121	94	88
15	140	121	93	88
Mean	132.9	117.8	88.8	85.9
S.D.	9.6	7.5	10.4	6.9
Paired t value	: -8.985		: -1.755	
p	: 0.0005		0.05	

DISCUSSION AND CONCLUSIONS

Antero-posterior cephalometric radiographs from 15 patients with PWS were compared to age- and sex-matched normal subjects. This study again showed that people with PWS are dolichocephalic, confirming the findings of Butler and Meaney (1987), although the value for one of our subjects (#1) was borderline. This borderline value was obtained from the youngest child in the study (4 years of age) and may suggest that this child had not grown sufficiently to fully express this trait. The dolichocephaly is not related to increase in facial height, since this measurement is not small (Schaedel et al, 1990), but rather to a narrowing of the cranial vault. The bifrontal diameter was consistently small in all patients with PWS.

It is also evident from this study that the face of people with PWS is not narrow overall. Bizygomatic width was statistically larger and bigonial and maximum cranial width were large in 11 of the 15 subjects. BIOD was also statistically large, and this is of particular interest because orbital hypertelorism has not been reported in the PWS. There also appears to be a gradual increase in BIOD with age, in contrast to the other measurements which appear to be relatively unaffected by age. A longitudinal study of growing subjects should be done to confirm or reject this finding.

Characteristics such as dolichocephaly, bifrontal diameter, BIOD, LOCP and bizygomatic width could be diagnostically useful in children and adults who present with other clinical features of the PWS. A larger study is needed to further delineate the differences in these measurements with growth.

REFERENCES

Aduss H, Przansky S, Miller M (1971) Interorbital distance in cleft lip and palate. Teratology 4:171-181

Butler M, Meaney J (1987) Anthropometric study of 38 individuals with Prader-Labhart-Willi syndrome. Am J Med Genet 26:445-455

Cassidy SB (1984) Prader-Willi syndrome. Current Problems in Pediatrics 14:5-8

Costavas M, Pruzansky S, Broadbent H (1982) Bony interorbital distance, head size, and level of the cribriform plate relative to orbital height: Normal standards for age and sex. J Cran Genet Develop Biol 2:5-18

Currarino G, Silverman F (1960) Orbital hypertelorism, arhinencephaly and trigonocephaly. Radiology 74:206-217

Farkas L (1981) Anthropometry of the head and face in medicine. Elserier North Holland, Inc., New York, p 8-12

Hall B. Smith D (1972) Prader-Willi syndrome. J Pediatr 81:286-293

Morin JD, Hill JC, Anderson MD, Grainger RM (1963) A study of growth in the orbital region. Am J Ophthal 56:895-901

Schaedel R, Poole A, Cassidy SB (1990) Cephalometric analysis of the Prader-Willi syndrome. Am J Med Genet 36:484-487

Wood GA (1950) Changes in width dimensions between certain teeth and facial points during human growth. Am J Ortho 36:676-700

SCOLIOSIS AND ITS TREATMENT IN THE PRADER-WILLI SYNDROME

Mel W. Jones
Lecturer
University Department of Orthopaedic Surgery
Glenfield General Hospital
Groby Road
Leicester, LE3 9QP
England

INTRODUCTION

Scoliosis in the Prader-Willi Syndrome:

Scoliosis in Prader-Willi syndrome (PWS) was first noted by Laurance (1967) who thought it was a common feature of the syndrome since it was also present in a number of patients referred by pediatric colleagues. It may to be due to hypotonia (Zellweger and Schneider, 1968). The reported prevalence rates vary widely from 7% (Zellweger and Schneider, 1968) to 97% (Holm and Laurnen, 1981). The latter emphasised that clinical examination alone is not reliable in detecting scoliosis in these short and obese children. In their series of 37, all of whom had spine radiographs, 36 (97%) had scoliosis. However, most were not severe. Only 5 (14%) had a scoliosis over 20 degrees, and only one patient underwent surgery.

Scoliosis is currently classified into early onset (presenting before 5 years of age) and late onset (presenting after the age of 5 years) (Leatherman and Dickson, 1988a). Most (60%) early onset idiopathic scoliosis is seen in boys, the curve being convex to the left in 90%. Late onset idiopathic scoliosis is much more common (90%) in girls, and 90% of curves are convex to the right (Apley and Solomon, 1982). The pattern of scoliosis in PWS appears to be similar to that in idiopathic scoliosis (Jones, 1990a).

NATO ASI Series, Vol. H 61
Prader-Willi Syndrome
Edited by S. B. Cassidy

Previous Reports of Scoliosis Surgery in PWS:

Prior to 1988, there were only 7 reports of patients with PWS who had undergone surgery for scoliosis. Five of these had uneventful surgery (Holm and Laurnen, 1981; Jaskulsky and Stone, 1987; Laurance et al, 1981; Gurd and Thompson, 1981; Soriano et al, 1988). Problems were encountered in 2 patients. One died post-operatively, but no details were given (Laurance et al, 1981). Another was described as having had "uneventful" and "unremarkable" surgery (Gurd and Thompson, 1981) but the laminae were too weak to hold a Harrington rod and fusion in situ had to be performed. Despite this, the authors concluded that, as far as the anesthesia and surgical care of these children was concerned, "there does not appear to be any major problems with the Prader-Willi children".

In 1989, a review of 3 patients from the Royal Liverpool Children's Hospital (Rees et al, 1989) reported major complications in all 3 patients. These included friable soft tissues, excessive bleeding, bone too soft to hold metal, pleural empyema, septicemia, cardiac arrest, prolonged post-operative ventilation, pneumothorax, CSF leak and meningitis. It was thought that the respiratory complications were related to performing a thoracotomy. Since a series of 3 is too small to allow meaningful conclusions to be drawn there was a clear need for a larger study, which was undertaken.

Patients and Methods:

As many patients as possible who had undergone scoliosis surgery, irrespective of whether complications occurred, were sought for the study. The organizations which were made aware of this study in a search for patients included the British Scoliosis Society, the Scoliosis Research Society, the PWS Association (UK) and the Scoliosis Association (UK). All authors of previously published reports were contacted and data on the 3 patients from the 1989 Liverpool study were included.

Whenever practical, case notes and x-rays were obtained for

study. As many cases were from abroad, this was not always possible and a 10 page questionnaire (Jones, 1990b) requesting specific data was sent to the clinicians concerned.

RESULTS

General:

A total of 33 patients were studied. There were 11 males and 22 females. The sources of the patients were the Scoliosis Research Society (21), the British Scoliosis Society (7), the PWS Association (UK) (3), and 2 cases previously published (Soriano et al, 1988; Gurd and Thompson, 1981). The mean age when the scoliosis was diagnosed was 7.4 years, ranging from 2 months to 17.5 years. Bracing had been tried in 24 patients prior to surgery. They had undergone surgery between 1969 and 1990 at a mean age of 12.4 years, ranging from 6.2 to 24.2 years. The countries where the patients were treated were USA (16), England (11), Sweden (2), Australia (2), France (1) and Holland (1).

Pre-Operative Pulmonary Function Tests:

A variety of tests were used in 12 patients, including tidal volume, peak expiratory flow rate, FEV1 and FVC. Abnormal results were obtained in 10 patients. In 3 of these, compliance was known to be poor but no comment about compliance was available for the other 7 cases. In the two cases which had a normal result, compliance was known to be good. Pulmonary function tests were not performed in 21 patients, often because of non-compliance.

Type of Surgery Undertaken:

Posterior spinal instrumentation was performed in 29 patients. This was preceded by anterior spinal growth arrest to remove the discs and adjacent epiphyses to obtain anterior fusion in 2 patients. The posterior spinal instrumentation used included Harrington rod (25), Cotrel-Dubousset (4) and Luque (2). Cotrel-Dubousset instrumentation is a recent development and involves the use of a series of hooks under the laminae and screws into the pedicles of vertebrae so that rods can be attached in a way that allows correction of the rotatory deformity in the spine as well as the deformity in the saggital plane. Luque instrumentation uses two parallel metal rods attached to the spine by a series of wires which pass under the laminae of the vertebra at several levels. Anterior Zielke instrumentation was used in 2 patients. This involves passing a screw into the front of several vertebral bodies. A flexible threaded metal rod is inserted through each screw. By tightening the nuts on the rod, correction of the scoliosis is obtained from the front of the spine. An additional costectomy was performed in 2 patients. Revision surgery was needed in 2 cases.

Complications of Surgery:

I. BLEEDING

Details of intra-operative blood loss and the surgeons' comment about this were available for 19 procedures. Blood loss was considered to be excessive in 18 of these. Therefore, excessive bleeding was encountered in at least 55%.

II. SOFT BONE AND TISSUE FRAGILITY

Soft bone was reported in 9 (27%) patients and fragile soft tissues in 3 (9%). In 4 patients, the procedure being performed had to be abandoned.

III. SEPSIS

This was encountered in 5 (15%) patients. There were 3 wound infections and another developed a CSF leak after an anterior growth arrest, with resultant meningitis. The fifth developed a persistent sinus after a pseudarthrosis was explored.

IV. RESPIRATORY PROBLEMS

These were encountered in 5 (15%) patients. One patient developed a pneumothorax after an anterior growth arrest and 4 patients needed prolonged intermittent positive pressure ventilation for respiratory insufficiency after surgery. One patient developed a pleural empyema after an attempt to insert a Harrington rod failed. She became septicemic, had a cardiac arrest, was successfully resuscitated and was ventilated for 22 days. Another was ventilated for 3 days. Details are not available about the others.

Other Surgery Performed:

Fourteen patients underwent surgery for other conditions. Often these were multiple and are shown in Table I. The only complication reported was a wound infection after an orchidopexy.

TABLE I. Other Surgical Procedures Performed

Strabismus correction	5
Orchidopexy	5
Hernia repair	3
Tonsillectomy	2
Dental	1
Skin ulcer excision	1
Leg lengthening	1
Liposuction	1
Pyloric stenosis	1
Quadriceps plasty	1
Circumcision	1
Open reduction and	1
Internal fixation of fractured femur	1

DISCUSSION

PWS occurs in 1:10,000-25,000 live births, and scoliosis appears to be an almost universal (97%) feature (Holm and Laurnen, 1981). Between 1969 and 1990 there were approximately 13 million live births in the United Kingdom (OPCS 1978, 1988). The number of children born with the PWS during this period should be approximately 500-1,300 although the exact number is not known. In this study there were 11 patients from the United Kingdom, and a further 4 are known to have had scoliosis surgery but the records were not available. Using the above figures, the proportion of children with PWS who have undergone scoliosis surgery is of the order of 1-3% in the United Kingdom. As the exact number of children born with PWS is not known with certainty, these figures are rough estimates. Indeed, they may be underestimates as it is possible that other children may have been operated upon of which I am unaware. However, it can be said that although scoliosis is very common, surgery is rarely needed. This is confirmed on a larger scale as the majority of interested scoliosis surgeons in the English speaking world were aware of this study and the total number of patients included is 33 despite a high level of interest and cooperation.

Pulmonary function tests could not be performed in some of the patients due to their inability to cooperate. In those who were tested, the results were often considered to be unreliable. In PWS, unless cooperation is known to be good, the results of pulmonary function tests should be viewed with caution and attention paid to exercise tolerance during play.

The excessive bleeding noted in 55% during surgery is a cause for concern. During the insertion of 14 Harrington rods and 1 Luque rod, the mean blood loss was 2145 ml, ranging from 625 to 5500 ml. This appears to be in excess of the expected blood loss of approximately a liter (Leatherman and Dickson, 1988b). A loss of 1030 mg during anterior growth arrest is twice that which is expected (Owen, 1990).

It is unclear why bleeding should be excessive. Analysis of age at the time of surgery was 12.4 in both those reported to have bled excessively and those who had not. Easy bruising has been reported in the syndrome (Holm, 1981), but no specific haemostatic defect has been identified. One of the patients in this study had a prolonged Ivy bleeding time of 750 seconds indicating thrombocytopenia, the upper limit of normal being 5 minutes (Houston et al, 1979). An important factor may be the increased tissue fragility noted in 3 (9%) patients in this study.

Soft bone was reported in 9 (27%) patients. In 4 of these this was a major problem resulting in the abandonment of the intended procedure. The mean age of those reported to have soft bone was 10.7 years compared to 13.1 years in those whom this had not been noted. The mean age of those reported to have fragile tissues was 11.8 years compared to 12.5 years in those whom this had not been noted. These differences were not statistically significant by student's t test.

The problem of soft bone may add some credibility to the claim that fracture after minimal trauma is said to occur in 5% of children with the syndrome. There may well be an underlying connective tissue disorder in the syndrome, although this is mere conjecture at present. This would be an explanation for the soft bone and fragile soft tissues and possibly the bleeding tendency. Another factor to be considered is the osteoporosis reported in the syndrome by Cassidy et al, (1985).

Sepsis was seen in 5 (15%) patients. This is high in comparison with spinal surgery in idiopathic scoliosis in which infection should be under 1% (Moe and Gustilo, 1964). It should be remembered that these patients have a tendency to pick at the skin which may lead to infection even without any surgery (Cassidy, 1987). However, wound infections were not found to be a problem after other types of surgery.

Respiratory problems were seen in 5 (15%) patients. Of a total of 5 anterior spinal procedures, there was one respiratory complication. Of 33 posterior procedures performed, there were 4 respiratory complications. Statistical analysis using the Chi squared chest did not show any statistically significant difference between these two groups. In the 1989 Liverpool study (Rees et al), it was postulated that a thoracotomy might be a factor in the development of respiratory complications but this does not appear to be true from the results of this larger study.

MANAGEMENT RECOMMENDATIONS

Screening:

All those managing children with PWS should be aware that

scoliosis is strongly associated with the syndrome. The children should be screened by physical examination initially. As short stature, obesity and lack of cooperation may make this difficult, an x-ray of the spine should be obtained if there is any doubt. If a scoliosis is found, these children should be under the care of a pediatric spinal surgeon. It is important not to over-irradiate these children and subsequent x-rays should be ordered after carefully considering the rate of progression of the scoliosis and the knowledge that very few of children with PWS will eventually require surgery.

<u>Conservative Management</u>:

Prevention is better than cure, and prone lying is known to have an inhibitory effect on early onset scoliosis, especially in a hypotonic infant (McMaster, 1983). Hypotonia, a prominent feature of PWS, is known to be a bad prognostic indicator (Mehta and Morel, 1979). It is, therefore, sensible to nurse PWS children prone from the moment of birth.

The scoliosis in PWS appears to be similar to that seen in idiopathic scoliosis. In progressive early onset scoliosis, there is evidence to support the use of serial elongation-rotation-flexion (EDF) casting (Mehta, 1977). However, the obesity may reduce the effectiveness of any cast applied.

In children with late onset scoliosis, the benefit of bracing has not been proven (Dickson, 1985). If the scoliosis is progressive, surgery is considered the only effective treatment (Miller, Machemson and Shultz, 1984; Dickson, 1985).

<u>Surgical Treatment</u>:

Should surgery be considered necessary, it is important for the surgeon to inform the parents of the potential problems of bleeding, soft bone, friable tissues, prolonged intermittent positive pressure ventilation post-operatively, and infection in addition to the other known risks of spinal surgery. The problem of soft bone needs particular emphasis as this may not allow the proposed procedure to be performed. Fusion in situ may have to be performed instead, with less correction than anticipated. The anesthetist involved must also be consulted and be aware of the potential problems. A decision to proceed with surgery must, therefore, be made after careful consideration.

If it is clear than an early onset scoliosis is progressive and likely to be severe despite casting, anterior spinal growth arrest and posterior spinal instrumentation should be offered and discussed with the parents. This is because there is a significant mortality and morbidity rate from cardiopulmonary compromise in untreated patients (Dickson and Archer, 1987). The complications likely to be encountered do not appear to depend on the child's age.

In late onset scoliosis, the problem is mainly a cosmetic one. In view of the potential problems, a decision to proceed with surgery must bear this in mind. If surgery is undertaken, posterior spinal instrumentation alone may be sufficient for scolioses less than 60 degrees. For more severe curves, an anterior spinal growth arrest should also be considered. An anterior spinal growth arrest does not appear to increase the respiratory complications that may be encountered. Whatever form of surgery is performed, hemostasis should be meticulous and the use of diathermy during dissection of the spine is to be strongly recommended.

Future Studies:

Although this study reports that excessive bleeding was encountered during scoliosis surgery, the reason for this is unclear and it does not appear to occur during minor surgery. One of the patients in this study had an abnormal Ivy bleeding time suggesting a platelet problem. I am not aware of any studies of the blood clotting mechanism in this syndrome. Such a study might reveal a defect which might be amenable to pharmacological manipulation.

Both bone and soft tissues were noted to be more fragile

than in normal children. Osteoporosis has also been reported. Qualitative and quantiative analysis of their connective tissue might help to elucidate the nature of PWS. Dual photon absorptiometry, computerized axial tomography and magnetic resonance imaging may be useful tools in such a study.

REFERENCES

Apley AG, Solomon L (1982) Apley's System of Orthopaedics and Fractures. London: Butterworth Scientific 222-223.

Cassidy SB (1987) Prader-Willi syndrome: Characteristics, management and etiology. Ala J Med Sci 24:169-175

Cassidy SB, Rubin KG, Mukaida CS (1985) Osteoporosis in Prader-Willi syndrome. Am J Hum Genet 37:A49

Dickson RA (1985) Conservative treatment for idiopathic scoliosis. J Bone Joint Surg 67B:176-181

Dickson RA, Archer IA (1987) Surgical treatment of late-onset idiopathic thoracic scoliosis. J Bone Joint Surg 69B:709-714

Gurd AR, Thompson TR (1981) Scoliosis in Prader-Willi syndrome. J Ped Ortho 1:317-320

Holm VA (1981) The diagnosis of Prader-Willi Syndrome. In Holm VA, Sulzbacher SH and Pipes PL (eds). Prader-Willi Syndrome. Baltimore, University Park Press. pp 27-44

Holm VA, Laurnen EL (1981) Prader-Willi syndrome and scoliosis. Devel Med and Child Neurol 23:192-201

Houston JC, Joiner CL, Trounce JR (1979) Disorders of the blood. In: Houston JC, Joiner CL, Trounce JR (eds). A Short Textbook of Medicine. Hodder & Stoughton, London, p 515

Jaskulsky SR, Stone NN (1987) Hypogonadism in Prader-Willi syndrome. Urology 29:207-108

Jones MW (1990a) Scoliosis surgery in the Prader-Willi syndrome. MChOrth Thesis, University of Liverpool, England, p 46

Jones MW (1990b) Scholiosis surgery in the Prader-Willi syndrome. MChOrth Thesis, University of Liverpool, England, pp 70-79

Laurance BM (1967) Hypotonia, mental retardation, obesity and cryptorchidism associated with dwarfism and diabetes in children. Arch Dis Child 42:126-139

Laurance BM, Brito A, Wilkinson J (1981) Prader-Willi syndrome after age 15 years. Arch Dis Child 56:181-186

Leatherman KD, Dickson RA (1988a) The management of spinal deformity. Wright, London, p 13

Leatherman KD, Dickson RA (1988b) The management of spinal defority. Wright, London, p 459

McMaster MJ (1983) Infantile idiopathic scoliosis: Can it be prevented? J Bone Joint Surg 65B:612-617

Mehta MH (1977) The natural history of infantile idiopathic scoliosis. In: Zorab PA (ed) Scoliosis: Proceedings of a fifth symposium held at the Cardiothoracic Institute, Brompton Hospital, London, September 21-22, 1976. Academic Press, London, p 103-122

Mehta MH, Morel G (1977) The non-operative treatment of infantile idopathic scoliosis. In: Zorab PA, Siegler D (eds) Scoliosis 1979: Based on the proceedings of the sixth symposium on scoliosis held at the Cardiothoracic Institute, Brompton Hospital, London, September 17-18, 1979. Academic Press, London, p 71-84

Miller JA, Nachemson AL, Schultz AB (1984) Effectiveness of braces in mild idiopathic scoliosis. Spine 9:632-635

Moe JH and Gustilo RB (1964) Treatment of scoliosis. J Bone Joint Surg 46A:293-321

OPCS (1978) Office of population censuses and surveys (UK). Birth Statistics 5:15

OPCS (1988) Office of population censuses and surveys (UK). Birth Statistics 17:21

Owen R (1990) Personal communication

Rees D, Jones MW, Owen R, Dorgan JC (1989) Scoliosis surgery in the Prader-Willi syndrome. J Bone Joint Surg 71B:685-688

Soriano RMG, Weisz I, Houghton GR (1988) Scoliosis in the Prader-Willi syndrome. Spine 13:209-211

Zellweger H, Schneider HJ (1968) Syndrome of hypotonia-hypomentia- hypogonadism-obesity (HHHO) or Prader-Willi syndrome. Am J Dis Child 115:588-598

PSYCHOLOGICAL PROFILE AND BEHAVIORAL CHARACTERISTICS IN THE PRADER-WILLI SYNDROME

L.M.G. Curfs
Research Department
Observation Center
De Hondsberg
Hondsberg 5
5062 jt Oisterwijk
The Netherlands

INTRODUCTION

The Prader-Willi syndrome is characterized by a number of clinical features including hypotonia, hyperphagia with childhood-onset obesity, hypogonadism, craniofacial abnormalities, a typical fat distribution and finely shaped hands and feet (Prader et al 1956; Cassidy, 1984; Butler et al, 1986; Greenswag and Alexander, 1988). This clinical picture, with the addition of delayed motor and language development, was first described in 1956 by Prader, Labhart and Willi and referred to as "Ein Syndrom von Adipositas, Kleinwuchs, Kryptorchidismus und Oligophrenie nach myatonieartigem Zustand im Neugeborenenalter".

To date, most studies of Prader-Willi Syndrome have been aimed at understanding diagnosis and etiology. Research into the psychological features has only just begun.

Current knowledge about the psychological characteristics of this syndrome will be reviewed, with special attention to the intellectual impairment and behavioral abnormalities, and the results of our investigations to further elucidate these abnormalities will be described.

PSYCHOMOTOR DEVELOPMENT AND SPEECH-LANGUAGE DEVELOPMENT

Infants with Prader-Willi syndrome (PWS) fall behind age mates in reaching important developmental milestones such as

NATO ASI Series, Vol. H 61
Prader-Willi Syndrome
Edited by S. B. Cassidy

rolling over, sitting up without support and walking (Cassidy, 1984; Zellweger, 1988). They rarely sit before the age of 1 year and usually do not walk before the age of 2 years. By the age of four years most of the children are able to walk, though their walking remains insecure and tottering. This is particularly true of the individuals having severe obesity problems. Fine motor skill development is also delayed.

The study of speech and language development is a relatively unexplored area. The obviously delayed speech-language development has been the subject of a very limited number of studies (e.g., Branson, 1981; Munson-Davis, 1988). Hypotonia has a negative influence on the muscles involved in speech production. Moreover, speech is affected by facial abnormalities which result in frequent saliva scuming, articulation problems and voice defects like nasality, decreased volume and poor fluency of speech (Dunn, 1968; Zellweger and Schneider, 1968; Branson, 1981; Siegel-Sandewitz and Shprintzen, 1982; Munson-Davis, 1988; Butler, 1990). The delay in speech-language development is intertwined with cognitive impairments.

BEHAVIORAL DEVELOPMENT

There is a remarkable similar pattern in behavior development in most patients. The first year of life is characterized by severe muscle hypotonia because of which sucking and swallowing problems are present, causing serious feeding problems. After two years of life, the infant overcomes the feeding problems because the muscle hypotonia diminishes. Then, at about two years of age, the child develops an insatiable appetite which leads to obesity if not controlled. This insatiable appetite becomes more noticeable as the child gets older. Everything revolves around food and eating for them. Other activities and interests tend to be pushed into the background. From now on, it is extremely important to pay close attention to the increase in weight. An important aspect of the guidance of parents and children from this point on is the teaching of healthy eating habits and weight control. Many of the behavioral problems in children with PWS are related to the abnormal appetite and the "search for food". There are no other specific behavioral problems in the first years.

Observation of patients and results of interrogation of parents and others such as teachers demonstrate that during the first years of life children with PWS are cheerful and cordial. They are experienced by their parents as quite dependent and rather passive without major behavioral problems.

After the age of four years, changes are noted. Mood changes, irritability and violent crying tantrums appear (Hall and Smith, 1972; Cassidy, 1984; Greenswag, 1987; Sulzbacher, 1988; Borghgraef et al, 1990). From childhood on the occurrence of impairments in initiating and maintaining relationships with peers are frequently described. However, there are no published empirical studies which compare data of different groups and which can clarify the specific nature of the social skill deficits. Whether they stem from cognitive deficits like learning or attention problems or are more directly related to the genetic abnormality remained unresolved.

At the age of puberty and thereafter a heightened vulnerability to emotional and the presence of personality disorders like negativism and depressive symptoms are described.

Parents, teachers and other care-personnel mention that, beginning at school age, lability in behavior and rapid alterations in mood are the most striking characteristics. Minor changes in the environment can result in sudden acts of inappropriate violence (Borghgraef et al, 1990). On the other hand, the PWS youngsters and adults are usually friendly and cooperative. (For reviews, see also Caldwell and Taylor, 1988; Greenswag and Alexander, 1988).

The absence of systematically gathered data on behavioral characteristics, including social-emotional functioning and personality characteristics, is striking. Still, personality and emotional problems are frequently mentioned by parents and professionals as one of the most troublesome areas, increasing in severity when individuals with PWS enter late childhood and adolescence. Cassidy (1984) among others reported that it is quite surprising to notice the lack of systematic studies on personality and psychological abnormalities. This remark still applies.

The Center for Human Genetics, University of Leuven, Belgium and the Research Department of the Observation Center, the Hondsberg, The Netherlands, conducted a study aimed at assessing cognitive and behavioral/emotional aspects of the PWS. The aim of one study (Curfs et al, 1991a) was to provide data on specific emotional and behavioral problems in a sample of youngsters with PWS aged 4 to 16 years.

In order to measure the existence and type of emotional and behavioral problems the Child Behavior Checklist (CBCL) was used. This is a questionnaire designed to report in a standardized way the behavior of children ages 4 to 16 years as assessed by their parents (Achenbach and Edelbrock, 1983). The items of the questionnaire describe a broad range of behavior problems that are of concern to parents as well as clinicians. Examples are: "Can't concentrate", "can't pay attention for long", "disobedient at home", "not liked by other children". The CBCL is normed for both sexes and for 3 age groups separately: 4-6; 6-11; 12-16 years.

The 90th percentile of the cumulative frequency distribution of total problem scores in normative samples is assigned as the cut off score above which children are scored in the clinical range and can be regarded as deviant.

Compared with the children of the same sex and age in the normative sample, the subjects with PWS showed significantly more problem behavior. The proportion of youngsters with PWS scoring in the clinical range was significantly greater than the proportion of deviant children in the normative samples. Of the total sample of patients under study, 81% were scored in the deviant range, whereas by definition the number corresponding with the 90th percentile in the normative sample is less than 10%.

Results of the study clearly confirm the presence and severity of problem behavior in individuals with PWS compared to agemates. Further investigation of the heightened vulnerability of PWS subjects is needed.

Accurate guidance and information for parents is important for avoiding the serious consequences for personality development from childhood on.

INTELLIGENCE AND COGNITIVE PROFILE

In the various studies of the psychological profile of PWS, consideration has been given to intellectual functioning. Results show that intellectual impairment is one of the characteristic clinical features of PWS. While most of the affected individuals are mentally retarded, studies also indicate a marked variation in intelligence.

For instance Peri et al (1984) presented a review of studies of intelligence between 1956 and 1980. It appears from this review that of the 155 PWS patients from 42 studies, most are mildly to moderately retarded. The average IQ was 57, ranging from 12 to 91 and a standard deviation of 15.

Greenswag (1987) reported results of 232 patients with PWS older than 16 years surveyed by questionnaire. Three percent of these patients were not mentally retarded. The classification of mental retardation of the other 97% was: borderline or mild mental retardation 63%, moderate mental retardation 31%, and severe mental retardation 3%.

In the Butler, et al (1986) study in which 39 children with the PWS were examined, mental retardation varied from moderate to borderline or mild mental retardation. The authors also presented a review of 7 other studies in the period of 1967 to 1984. It appeared from these studies that most of the patients with PWS were mentally retarded and belonged to the category mild mental retardation.

In the Curfs and Fryns study (1991), the data was summarized from 57 selected studied, covering a total of 575 patients with PWS, all mentioning IQ testing and subclassification according to the American Association on Mental Deficiency guidelines (Grossman, 1983). From that review it could be concluded that (1) approximately 5% of the patients were not mentally retarded and have IQ's compatible with normal education (IQ's above 85); (2) the majority of patients belonged to the categories borderline (27%), mild (34%) and moderate (27%) mental retardation; (3) it was noteworthy that only a minority of the patients was at the severely retarded level (5%) and functioning at the profoundly retarded level was exceptional (<1%); (4) there appeared to be no clear differences between females and males in cognitive level; (5) nor were there differences in the distribution of intelligence of children compared to adults.

COGNITIVE PROFILE

Little is known about the broader spectrum of cognitive functions and the specific structure of cognitive abilities. For instance, a comparison between verbal and performance intelligence is seldom mentioned in literature. The question has not been explored of whether individuals with PWS show specific cognitive strengths and weaknesses by comparing subtest scaled scores to mean scaled scores of intelligence testing of that particular individual.

This was the aim of the study by Curfs et al (1991). Results indicate that cognitive strengths were more apparent than cognitive weaknesses. The group of 26 individuals under study showed a flat profile with relatively few valleys. Peaks were clearly evident, especially in the performance scale. The frequent high score on Block Design was impressive, reflecting an ability to recognize and evaluate figure relations.

Analysis of subtest scores and comparison of differences between verbal IQ and performance IQ showed that individuals with PWS did better on visual motor discrimination skills than on auditory verbal processing skills. These results are in accordance with the studies of Gabel et al. (1986) and Taylor (1988), who also reported strengths in visual perception and organization.

The issue of a specific cognitive profile for the PWS remains a topic of controversy. Further studies are warranted, particularly a comparison of patients with PWS versus patients with other chromosomal disorders and with the group of learning disabled individuals.

FOOD-RELATED BEHAVIOR AND NUTRITIONAL MANAGEMENT

A cardinal symptom of the PWS is the obesity. If the child with PWS is not given proper guidance and care with regard to eating habits and weight control from an early age, obesity problems occur and can be life threathening. To overcome overweight problems is not an easy task for the child, parents and other involved caregivers. Only in the beginning of the 1970s, after the publication of Pipes and Holm (1973) aimed at weight control and nutritional management, did more publications appear which were directed towards influencing obesity. There have been approximately 60 studies with regard to weight control (reviewed by Rasenberg et al., in press). Nineteen of these studies mentioned the exceptional eating behavior and obesity. Intervention strategies directed at controlling the obesity and inducing weight loss is the topic of the other 40 studies. Different approaches have been tried to prevent patients with PWS from gaining excessive weight, such as dietary management, appetite suppressing drugs, surgery such as gastric bypass, gastroplasty, truncal vagotomy, behavioral management, and combinations of these. Assessment of why some of the eating control programs fail and why some programs are successful is an important area for future research.

Important elements of successful weight control programs are (1) attention to the complex emotional implications of dietary management, since neglecting parent attitudes, motivation and feelings in this respect is disastrous; (2) emphasis on self-control-mechanisms (personal commitment to dieting) and the development of a positive self-image in the participant; (3) built-in elements of a followup program; (4) accommodating the program and instructions to the intelligence level of the individuals with PWS; (5) functional analysis of the eating behavior problems and environmental controls; and, (6) applying a program based upon insight in the mechanisms of appetite control and satiety of the particular participant. Programs with exclusive concentration on diet prescription without regard to the above mentioned aspects are unlikely to be successful.

DISCUSSION

Of all psychological aspects, most attention has been directed towards cognitive functioning and food related behaviors. Individuals with PWS have significant disabilities in learning. However, intelligence levels of both males and females show a wide variation. The majority of the patients fall in the range of borderline to moderate mental retardation. Thus far there have been few systematic studies on cognitive profile in patients with PWS. Information processing involving the auditory modality is relatively weak in comparison to visual processing. There has been little systematic research aimed at personality, social and emotional development. This relatively unexplored area is very important, because it is quite clear that specific characteristics of the syndrome, such as recognizable clinical phenotype, the cognitive impairments, the language and speech difficulties, the social skills deficits, the poor self-image and the aberrant behaviors, make the individual with PWS very vulnerable in society.

Attention to the broader spectrum of cognitive functioning by systematic assessment of specific cognitive skills such as perception, memory and language is necessary. Longitudinal research is especially warranted.

Although many questions with regard to the syndrome have remained unresolved thus far, important progress in some areas is being made. Prevention of the dramatic obesity which in turn leads to severe physical problems and psychological burdens in individuals with PWS is possible by giving full and accurate information about the syndrome to parents and professionals and by providing appropriate guidance.

It is very important to have any early diagnosis of this Multiple Congenital Anomaly Syndrome. It is a "conditio sine qua non" for the prevention of major complications in children and adults with PWS. Butler et al (1986) found the mean age of diagnosis for their sample of 29 PWS patients to be 7 years and 2 months, with a range of 6 weeks to 32 years. The mean age of diagnosis of the 55 PWS patients participating in our multicenter Dutch-Belgium study was 7 years and 6 months. It is obvious that the time of diagnosis should be much earlier in order to obtain maximum benefit from intervention programs.

REFERENCES

Achenbach TM, and Edelbrock CS (1983) Manual for the Child Behavior Checklist and Revised Child Profile. Burlington VT, Department of Psychiatry University of Vermont

Borghgraef M, Fryns JP, and Van den Berghe H (1990) Psychological profile and behavioral characteristics in 12 patients with Prader-Willi syndrome. Genetic Counseling 38:141-150

Branson C (1981) Speech and language characteristics of children with Prader-Willi syndrome. In: Holm VA, Sulzbacher SJ and Pipes PL (eds) The Prader-Willi Syndrome. University Park Press, Baltimore, pp 179-183

Butler MG (1990) Prader-Willi syndrome: Current understanding of cause and diagnosis. Am J Med Genet 35:319-332

Butler MG, Meaney FJ, Palmer CG (1986) Clinical and cytogenetic survey of 39 individuals with Prader-Labhart-Willi syndrome. Am J Med Genet 23:793-809

Caldwell ML and Taylor RL (1988) Prader-Willi syndrome: Selected Research and Management Issues. Springer-Verlag, New York

Cassidy SB (1984) Prader-Willi syndrome. Curr Prob in Pediatr 14:1-55

Curfs LMG, Wiegers AM, Sommers JRM, Borghgraef M and Fryns JP (1991) Strengths and weaknesses in the cognitive profile of youngsters with Prader-Willi syndrome. Clin Genet (in press)

Curfs LMG, Verhulst FC and Fryns JP (1991) Behavioural and emotional problems in youngsters with Prader-Willi syndrome. Genet Counsel (in press)

Curfs LMG and Fryns JP (1991) Prader-Willi syndrome: A review with special attention to the cognitive and behaviour profile. Psychosocial Aspects of Genetics. Wiley & Sons, New York (in press)

Dunn HG (1968) The Prader-Labhart-Willi syndrome: Review of the literature and report of nine cases. Acta Paediatr Scandinavica, supplement 186:1-38

Gabel S, Tarter RE, Gavaler J, Golden WL, Hegedus AM, Maier B (1986) Neuropsychological capacity of Prader-Willi children: General and specific aspects of impairment. Applied Research in Mental Retardation 7:459-466

Greenswag LR and Alexander RC (1988) Management of Prader-Willi Syndrome. Springer-Verlag, New York

Greenswag LR (1987) Adults with Prader-Willi syndrome: A survey of 232 cases. Devel Med Child Neurol 29:145-152

Grossman HJ (ed) (1983) Classification in Mental Retardation. American Association on Mental Deficiency, Washington, D.C.

Hall BD and Smith DW (1972) Prader-Willi syndrome. J Pediatr 81:286-293

Munson-Davis JA (1988) Speech and language development. In:Greenswag LR and Alexander RC (eds) Management of Prader-Willi Syndrome. Springer-Verlag, New York

Peri G, Molinari E and DiBlasio P (1984) Psychological observations on patients with the PWS. Acta Medica Auxol 16(1):29-43

Pipes PL and Holm VA (1973) Weight control of children with Prader-Willi syndrome. J Am Diet Assoc 62:520-524

Prader A, Labhart A and Willi H (1956) Ein Syndrom von Adipositas, Kleinwuchs, Kryptorchidismus und Oligophrenie nach myatonieartigem zustand im Neugeborenenalter. Schweiz Med Wochenschr 86:1260-1261

Rasenberg SAAJ, Wiegers AM, Curfs LMG, and Fryns JP. The Prader-Willi syundrome: A review of weight control and nutritional management programs. In Curfs LMG and Fryns JP (eds), Food-Related Behavior and Nutritional Management (in press)

Siegel-Sandewitz V and Shprintzen RJ (1982) The relationship of communication disorders to syndrome identification. J Speech and Hearing Dis 47(1):338-354

Sulzbacher S (1988) Psychological and behavioural management. In: Greenswag LR and Alexander RC (eds) Management of Prader-Willi Syndrome Springer-Verlag, New York pp 99-112

Taylor RL (1988) Cognitive and behavioural characteristics. In: Caldwell ML, Taylor RL (eds) Prader-Willi Syndrome: Selected Research and Management Issues. Springer-Verlag, New York pp 29-43

Zellweger H and Schneider H (1968) Syndrome of hypotonia, hypomentia, hypogonadism-obesity (HHMO) or Prader-Willi Syndrome. Am J Dis Child 115:588-598

Zellweger H (1988) Differential diagnosis in Prader-Willi syndrome. In: Greenswag LR and Alexander RC (eds) Management of Prader-Willi syndrome. Springer-Verlag, New York

THE USE OF PSYCHOTROPIC MEDICATIONS IN PERSONS WITH PRADER-WILLI SYNDROME

Barbara Y. Whitman, Ph.D.
Cardinal Glennon Hospital, KC Center
1465 S. Grand Boulevard
St. Louis, Missouri 63104
U.S.A.

Louise R. Greenswag, Ph.D.; Department of Pediatrics, University of Iowa Hospital and Clinics; Iowa City, Iowa, U.S.A.

INTRODUCTION

Prader-Willi syndrome (PWS) is an unusual and multifacted birth defect characterized by severe hypotonia, hypogenitalism, delayed development, and mental retardation (Prader et al, 1956; Zellweger and Schneider, 1968; Zellweger 1988). Hyperphagia becomes evident early in childhood and is accompanied by predisposition to increasingly unstable emotions and behavior problems that intensify over time (Sulzbacher, 1988). If the insatiable appetite is left unmanaged, affected individuals may become morbidly obese and die of complications of obesity. A no less important component of the syndrome is the evolution over time of inappropriate and often bizarre behaviors that escalate in frequency and intensity with age, necessitating ongoing management and intervention (Greenswag, 1987). Anecdotal reports of chemical management of these behaviors, accompanied by conflicting opinions regarding their appropriateness and effectiveness, are emerging. No data exists to guide clinicians' choice of medication, dosage, potential side effects, and time frame for administration. Moreover, information comparing the effectiveness of psychotropic medications with non-drug behavioral intervention is not available.

This report summarizes the first of a two-part study of the history, current use, and impact of mood altering and behavior change medications as a method of behavior management in adolescents and adults with PWS. This investigation posed several questions. What proportion of persons with PWS are managed with

NATO ASI Series, Vol. H 61
Prader-Willi Syndrome
Edited by S. B. Cassidy

psychotropic medications? Is there a relationship between living arrangement and administration of behavior control drugs? Does a relationship exist between weight and a need for behavior control? And, finally, is there an association between weight, the extent to which the living environment is controlled, and drug interventions? For purposes of this study, the term "drugs" is restricted to that class of medications termed psychotropic/psychoactive.

METHODS

Sample: The history and current use of psychotropic medications was investigated in a sample of 114 persons with PWS age 15 and older. It consisted of 56 males and 58 females who lived either at home with their families, or in one of three out-of-home settings: (1) dedicated residential programs comprised entirely of persons with PWS, (2) integrated group homes which served persons with PWS and other mentally retarded adults, (3) supervised apartment living. Those living at home (N=34) were randomly selected from a national list available for research through the Prader-Willi Syndrome Association in the U.S.A.. The out-of-home sample (N=80), was drawn from a published list of agencies offering residential services and was geographically stratified.

Instrument: Data was gathered by personal interview and responses to a mailed survey adapted from a National Institute of Mental Health Epidemiological Catchment Area (NIMH-ECA) instrument. In addition to inquiries regarding psychotropic drugs, two questions addressed the use of appetite suppressants and medications for gastrointestinal distress. Open-ended questions encouraged collection of anecdotal reports regarding the effectiveness of the identified drugs. Specific categories of medications associated with the presence of anxiety, depression, emotions, sleep disturbances, and behaviors were examined in relation to sex, age, current weights, and residential status. The data reported here is restricted to that portion of the questionnaire pertaining to psychotropic drugs.

Procedures: The administrators of selected group homes and the parents/guardians of record in the "living at home" sample received a letter of introduction from the investigators that explained the goals and objectives of the study. It contained a stamped, self-addressed postcard which, when returned, indicated consent to participate. Returned cards also indicated age and sex of the person with PWS, contact phone numbers, and a time most convenient for the investigator to conduct phone interviews with the respondents.

RESULTS

Table I lists the age, sex and size characteristics of the 114 patients with PWS included in the study.

Table I. Sample Characteristics

	Total	Male	Female
N	114	56	58
Age	27.3 (15-48)	26 (16-47)	28.6 (15-48)
Wt (kg)	65.52 (30-117)	67.2 (30-117)	63.9 (34-91)
Ht (cm)	148.4 (124-166)	149.5 (131-166)	147.2(124-156)

Within the total sample 34 (30%) persons lived at home (17 males and 17 females), 68 (60%) persons lived in dedicated group homes (33 males and 35 females), 7 (6%) persons lived in integrated group homes (4 males and 3 females), and 5 (4%) persons lived in supervised apartments (2males and 3 females).

The data indicated that 42 (37%) of the total sample had at some point been given psychotropic drugs; no significant differences between sexes were noted. Of the 42 cases to whom drugs were given, use in 6 (5%) was subsequently withdrawn due either to toxicity or ineffectiveness, leaving a total of 36 (32%)

persons currently receiving psychotropic medications (17 females and 19 males).

The 102 persons who lived either in their family's home or in dedicated group homes were more closely examined. Weight data was missing for one female, thus reducing the total number to 101.

In keeping with this study's primary concern of potentially excessive use of psychotropic medications as a means of behavioral control in group home environments, the differential use of drugs between those living in group homes and those in their family's home was compared. Twenty-one (31%) of all 67 group home residents and 12 (35%) of the 34 persons living at home were receiving medication. A Chi square analysis indicated no significant differences between these two settings.

Looking at differences between sexes, while holding place of residence constant for the 101 subjects, showed that 18 (36%) of the 50 males and 15 (29%) of the 51 females were receiving drugs. Chi square analysis again yielded no significant difference.

Anecdotal evidence suggests that individuals with lower weights often exhibit more behavior problems. Several analyses were conducted to ascertain if indeed the thinner persons were more often medicated. To standardize weights for analyses, all weights were converted to standard deviations from expected weight for height. Current management by some clinicians includes a leeway of up to 4 standard deviations above expected weight for height before obesity intervention is considered crucial unless other weight related health issues are pressing, due to the aberrant proportion of lean body mass in PWS (Cassidy, 1991). Therefore, all analyses were conducted using a cut-off point of greater or less than 4 standard deviations. When comparing sex and weight while holding place of residence constant, no differences in the frequencies of drug administration were found (Table II). When those above and below 4 standard deviations in weight were compared by place of residence (Table III), a Chi square of $p < .05$ indicated that more cases living in group homes weighed less than 4 s.d.

above expected weight for height. Furthermore, when those cases given medication were compared with those off medication by weight (Table IV) a significant Chi square ($p < .05$) suggests that those weighing less than 4 s.d. above expected weight for height were more frequently on drugs. Additional examination within each setting by sex, weight and medication use indicated that the number of females given drugs in residential settings whose weights were less than 4 s.d. above weight for height far exceeded expected frequencies.

Table II. Comparison of Drug Use by Weight and Sex

	<4 s.d.	>4 s.d.
Males	30	20
Females	30	21

$Chi^2 = .01$
O.R. = 1.05

Table III. Comparison of Drug Use By Weight and Living Arrangement

	<4 s.d.	>4 s.d.
Dedicated Group Home	45	22
Family Home	15	19

$Chi^2 = 4.9$
O.R. = .39
$p < .05$

Table IV. Comparison of Drug Use by Weight

	<4 s.d.	>4 s.d.
On Drugs	24	9
Not On Drugs	36	32

$Chi^2 = 4.3$
O.R. = 2.37
$p < .05$

DISCUSSION

Historically, the advent of psychotropic medications early in the 1960's has played a revolutionary role in the care of the mentally ill, the retarded and the developmentally disabled. Over time, use of these medications has had both a positive and a negative impact on the delivery of services to the mentally retarded population in general and, as noted here, to persons with PWS. Psychotropic drugs prescribed by qualified professionals and in combination with other treatments have, for many, been a godsend. However, indiscriminate use and abuse, particularly in the mentally retarded and developmentally disabled population, is now well documented (Minnesota Department of Public Welfare, 1980). While the real purpose of the chemical tools is to reduce anxiety and enhance receptivity to psychotherapeutic healing and growth, too frequently these medications are employed as end treatments to control behaviors; unruly patients become less troublesome and difficult, case management is less of a burden, and the work of caregivers is easier.

The behavioral difficulties which tend to escalate over time in persons with PWS result in poor psychosocial adaptation. Unlike their normal peers, persons with PWS have difficulty developing compensating emotional and behavioral responses to overcome emotional vulnerabilities. Behavioral volatility and labile moods are a direct consequence of the presence of the syndrome itself, rather than an outcome of poor management. Admittedly, a small number have significant psychiatric difficulties; most, however, do not. Prior reports concerning the use of psychotropic medications in individuals with PWS have been both anecdotal and conflicting, suggesting widespread misuse to control characteristic aberrant behaviors.

While the results of this study to not support the assertion that widespread misuse is occurring, they do raise a number of significant issues that require further study. For example, the data indicate that less obese persons with PWS appear to be receiving psychotropic medications more frequently than their heavier peers. While it might be tempting to suggest that these drugs may account for some of the weight loss, two factors argue against such a conclusion (Figure 1).

First, psychotropic drugs notoriously precipitate rapid and excessive weight gain. By contrast, these data indicate that persons living in dedicated group homes consistently weighed less than those living elsewhere whether or not they were medicated. There was a significantly increased administration of drugs to less obese persons in settings where weight was strictly managed.

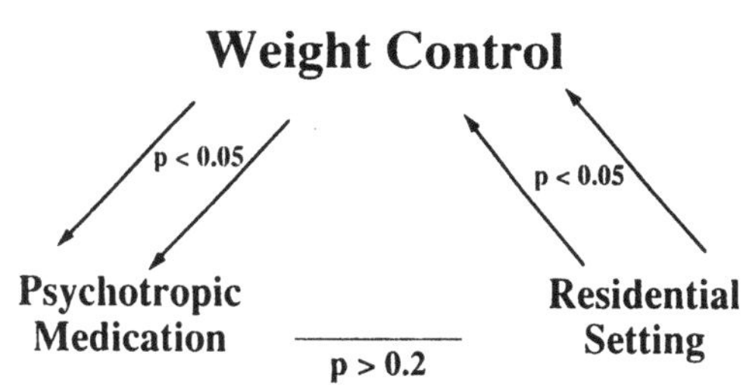

Figure 1. Causal pathway 1.

Second, while there was no overall significant difference between sexes in the administration of these drugs, slimmer females in group home settings were significantly more often given medications. Thus, the data suggest the possibility that the causal chain begins with weight reduction which then may lead to an increase in the quantity and severity of behavior problems. These, in turn, may be more frequently managed through chemical rather than programmatic means in group home settings (Figure 2).

Confirming previous clinical observation, the present data suggest that persons with PWS who are at more normal weight for height may have increased behavior difficulties. Previous data has suggested that these same persons also have increased emotional lability. Questions arise as to whether the association between thinness and more behavioral problems is a real phenomenon or simply a by-product of weight reduction. Do persons with PWS who have been well managed (and thus have never become massively obese) develop similar behavior lability as they reach adolescence and beyond? Is management of behaviors by chemical means more effective than non-drug interventions? Are drugs more or less effective over time? Why are drugs used significantly more often with females? Is this last question a reflection of more severe behaviors in females or completely different dynamics? What alternatives to drug interventions are available? Clearly, much study is yet required to answer these questions.

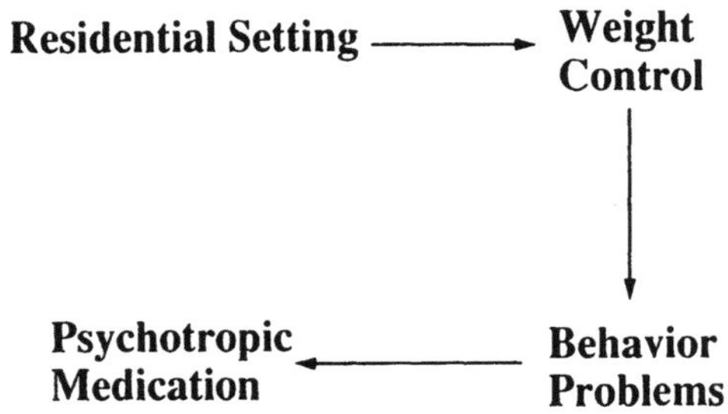

Figure 2. Causal pathway 2.

REFERENCES

Cassidy SB (1991) Personal communication

Greenswag LR (1987) Adults with Prader-Willi syndrome: A survey of 232 cases. Dev Med Child Neurol 29:145-152

Minnesota Department of Public Welfare (1980) The Use of Medications in Controlling the Behavior of the Mentally Retarded. Proceedings of the conference on the Use of Medications in Controlling the Behavior of the Mentally Retarded. Sponsored by the Department of Public Welfare, State of Minnesota, September 22-24, 1980

Prader A, Labhart A and Willi H (1956) Ein Syndrom von Adipositas, Kleinwuchs, Kryptochismus und Oligophrenie nach Myatonieartigem Zustand im Neugeborenalter. Schweiz Med Wochenschr 86:1260-1261

Sulzbacher S (1988) Psychological and behavioral management. In: Greenswag LR and Alexander RC (eds), Management of Prader-Willi syndrome. Springer-Verlag, New York, pp 99-122

Whitman BY and Accardo P (1987) Emotional symptoms in Prader-Willi adolescents. Am J Med Genet 4(1):636-638

Zellweger H and Schneider H (1968) Syndrome of hypotonia-hypomentia-hypogonadism-obesity (HHHO) or Prader-Willi syndrome. Am J Dis Child 115:588-598

Zellweger H (1988) Differential diagnosis in Prader-Willi syndrome. In: Greenswag LR and Alexander RC (eds) Management of Prader-Willi syndrome, Springer-Verlag, New York pp 15-22

ANGELMAN SYNDROME IN THE ADOLESCENT AND YOUNG ADULT

Jill Clayton-Smith
Department of Medical Genetics
St. Mary's Hospital
Whitworth Park
Manchester M13 0JH
England

INTRODUCTION

Generally speaking, most Angelman syndrome (AS) patients have been referred initially to a pediatrician or pediatric neurologist because of global developmental delay, particularly delayed motor milestones. Over the last five years there have been many case reports of children with AS, and this has led to an increased awareness of the condition and to diagnosis at an earlier age. Some of the most common questions asked by parents concern what the future holds for their child, and with a paucity of reports of older patients in the literature, information is scarce. This prompted a study of adolescents and young adults with AS in an attempt to assess the extent to which they are able to develop skills essential for daily living. Although the information obtained is useful, it is important to remember that the newly diagnosed children of today have access to much more physical, occupational and behavioral therapy, and this will almost certainly reflect favorably on their future achievements as adults compared to this study group.

METHODS

Seventeen Angelman syndrome (AS) patients aged between 10 and 24 years have been studied as part of a larger clinical study of AS in the United Kingdom. This work has been funded by Action Research for the Crippled Child, a charity organization in the United Kingdom. Patients have been visited in their own homes

NATO ASI Series, Vol. H 61
Prader-Willi Syndrome
Edited by S. B. Cassidy

and examined individually by the author. There were 5 male and 12 female patients in the group, and age at diagnosis ranged from 5 to 20 years. Only 2 patients were in institutional care.

RESULTS

The facial phenotype of AS is an evolving one. The features are unremarkable in the baby and infant, when the diagnosis is more likely to be made because of jerky movements and EEG abnormalities. As the child grows the characteristic features of prognathism, wide smiling mouth and mid-face retrusion become more apparent and those persist into adult life, the jaw becoming even more prominent as the face elongates. The facial features do not coarsen, and most patients look remarkably youthful for their age.

Eleven patients were ambulatory at the time of examination. Two had become non-ambulatory and had developed contractures of their lower limbs. Four patients had significant scoliosis. These were of the infantile type and were progressive, especially once puberty was reached and on becoming wheelchair bound. Two patients had required surgery for their scoliosis, and spinal fixation had been recommended in the third.

Although 12/17 patients had some speech and the maximum number of words was 5, the main means of communication was by sign language (1-50 Makaton signs) or primitive gestures. Four patients had neither speech nor signs.

Sixteen of 17 patients had had seizures, but these had decreased in frequency in all patients and ceased in 6. The EEG appearances which are very characteristic in AS (Boyd et al, 1988) become less apparent with age, and more normal rhythms developed. The diagnosis might be disputed on EEG criteria, therefore, if the first EEG is only performed at this age.

Eleven of 12 females were menstruating normally with a

regular cycle. Age at menarche ranged from 10-16 years. There was an increased likelihood of seizures perimenstrually.

Fourteen of 17 patients were toilet trained by day, but only 5 were dry at night. Fourteen patients fed themselves. Twelve used a knife and fork with good effect. No patient could dress unaided but 9/17 dressed with help.

There was an improvement with age in hyperactivity, sleep patterns and the seizure disorder. However, as Angelman individuals grow older they may develop obesity and immobility due to the development of contractures and the progression of scoliosis. The over-affectionate nature which is one of the hallmarks of this condition (Angelman, 1965), and is generally acceptable in the younger child, persists into adulthood when it can present a management problem. These patients cannot acquire skills to the extent of their being capable of independent living, but are able to learn some skills as aids to daily living.

REFERENCES

Angelman H (1965) Puppet children. Dev Med Child Neurol 7:681-688

Boyd SG, Harden A, Patton MA (1988) The EEG in early diagnosis of the Angelman (happy puppet) syndrome. Eur J Paediatr 147:508-513

CLINICAL FINDINGS IN INDIVIDUALS WITH ANGELMAN SYNDROME WITHOUT A MOLECULAR DELETION OR UNIPARENTAL DISOMY

Roberto Zori, M.D.
Department of Pediatrics
Division of Genetics
Box J-296
University of Florida
Gainesville, Florida
USA

R. Nicholls[1,2], D.J. Driscoll[1] and C.A. Williams[1], Division of Genetics, Department of Pediatrics[1], and Department of Neuroscience[2], University of Florida, Gainesville, Florida, USA

INTRODUCTION

About 60% of people with Angelman syndrome (AS) have a cytogenetic deletion of chromosome region 15q11-q13. The deleted chromosome has consistently been demonstrated to be of maternal origin (Magenis et al, 1987; Knoll et al, 1989; Pembrey et al, 1989; Williams et al, 1990). The Prader-Willi syndrome (PWS) is often associated with a deletion of the same chromosomal area but with paternal origin of the deleted chromosome (Knoll et al, 1989; Butler et al, 1983; Butler et al, 1986). Recently, maternal uniparental disomy has been found in some patients with PWS (Nicholls et al, 1989b) and a report of two patients with AS and probably disomy has been published (Malcolm et al, 1991). The purpose of the present study was to perform molecular analysis of 15(q11-q13) in lymphocytes of individuals with AS and then review their clinical findings to determine if deletion negative patients could be clinically distinguished from deletion positive ones. Clinical differences would imply that genetic heterogeneity exists in Angelman syndrome.

NATO ASI Series, Vol. H 61
Prader-Willi Syndrome
Edited by S. B. Cassidy

METHODS

Twenty-six individuals with AS were identified through the Angelman Research Group. This group is a university referral and evaluation project as well as a family support effort. The age of individuals with AS in our study ranged from 3-20 years, all individuals were caucasian, and the male:female ratio was 2:3. Two clinical geneticists were involved in making the diagnosis using the following criteria: 1) microbrachycephaly; 2) seizures of varied type; 3) awkward, stiff and jerky gait with flexed upper arms; 4) absent speech; 5) characteristic facial appearance with prognathism, macrostomia, protruding tongue, and frequent laughter; and 6) abnormal EEG with diffuse, markedly abnormal sharp and slow waves of the 2-4 Hz type (Boyd et al, 1988). Because we have felt confident making the diagnosis of AS clinically, we generally have not used EEG diagnostically.

The individual patients were studied by standard molecular methods (RFLP and dosage) (Nicholls et al, 1989b) using the following DNA probes: 3-21, 34, 189-1, IR4-34 and IR10-1 looking for the presence or absence of a 15q deletion. Two of the patients have previously been studied by Knoll et al (1991). These DNA sequences span the area that is known to be deleted in those patients with AS who have a deletion (Knoll et al, 1990). The patients' clinical histories were then reviewed individually. The developmental and medical history was reviewed and a physical exam was performed on each patient.

RESULTS

All patients had either microcephaly or relatively small heads in comparison with their bodies and with their parents. All of the patients had an awkward, stiff gait with uplifted arms. None had any speech and all had the facial and behavioral findings typical for AS. Twenty-one patients were found to have a deletion of chromosome 15 and no cases of uniparental disomy were found, but no deletion was detectable in 5 individuals. Most deletion patients and all nondeletion patients were described by their parents as being happy, with outbursts of laughter. All exhibited behavior consistent wiht AS during examination by one of the authors. The children were active, happy and inquisitive. All the children were healthy, and any physical findings or medical problems could be attributed to AS. Table I compares selected features found in the five cases with features of deletion positive Angelman syndrome individuals. No features appeared significantly more often in any group by Fisher Exact test.

Table 1. Presence of clinical findings in five Angelman syndrome (AS) individuals without a cytogenetic or molecular 15q11-13 region deletion as compared to a cohort of AS individuals with deletion

	AS without Deletion					AS with Deletion	
	1	2	3	4	5	n	%
Height < 5%	–	–	–	–	–	7/16	43
Weight < 5%	–	–	–	–	–	4/16	25
Head Circumference < 5%	+	+	–	+	–	14/16	88
Hand length < 5%	–	+	+	–	+	4/15	27
Foot length < 5%	–	+	+	–	+	5/15	33
Hypopigmentation	–	–	+	–	+	9/16	56
Protruding tongue	+	+	–	+	–	13/16	81
Prognathia	+	+	+	+	+	12/16	77
Wide-spaced teeth	+	–	+	+	+	8/16	50
Jerky gait	+	+	+	+	+	15/16	94
Uplifted arms	–	+	+	+	+	11/16	69
Pronated feet	+	–	+	–	–	8/16	50
Absent speech	+	+	+	+	+	16/16	100
Frequent laughter	+	+	+	+	+	14/16	87
Seizures	+	–	–	–	+	14/16	87

<u>Case 1:</u> J.S. was a 20-year-old female. Her perinatal history was unremarkable and her birth weight, height and head circumference were normal. There were some feeding difficulties in infancy requiring formula manipulation. She walked with support at about 10 months and delay was first noted at 9-12 months. She had erratic sleeping habits, getting up multiple times during the night. She understood simple commands but had no speech. Seizures

appeared at 1-1/2 years of age and these were severe, requiring treatment with adrenocorticotropic hormone (ACTH). They disappeared at about 2 years of age and she continued to be seizure free. Her EEG had low to moderate 4-6 cycle/second rhythm superimposed on an irregularly occurring, moderate high voltage spike and wave/poly spike and slow wave activity. The result of a brain computed tomography (CT) was normal. Her height was at the 50th centile, her weight was 25th centile and her head circumference was <5th centile. She could not walk alone, but when assisted had an ataxic, jerky gait. Her face was characteristic, with a large jaw and widely spaced teeth, and pronated feet. She had no obvious hypopigmentation.

Case 2: K.S. was a 5 1/2-year-old female. Perinatal history was unremarkable with a normal birth weight, height and head circumference. She had failure to thrive neonatally and was a restless sleeper. Developmental delay was first suspected at 3 months of age. She walked first at 2 3/4 years. She had never learned to talk but communicated in sign language with about 10-15 signs. She had never had seizures. The results of an EEG and CT scan were normal. Her height and weight were at the 5th centile and her head circumference was <5th centile. She walked with a stiff, unsteady gait with uplifted arms. Her tone was decreased. Her hands were small (<5th centile) and her feet were at the 5th centile. She was not hypopigmented. She had nystagmus.

Case 3: A.M. was a 6-year-old male. He had a normal perinatal history with a normal birth height, weight and head circumference. He did not nurse well and developmental delay was first noted at 3-4 months. Sleeping problems were not significant. He has never had seizures and an EEG and brain CT scan were reported as normal. His height and weight were >97th centile and his head circumference was at the 5th centile. His foot size was <5th centile and his hand size was normal. He walked with an ataxic, jerky gait and he had a large jaw. He had lightly pigmented skin, irides and hair. Of note, his father's height was >97th centile and his mother's was at the 75th centile.

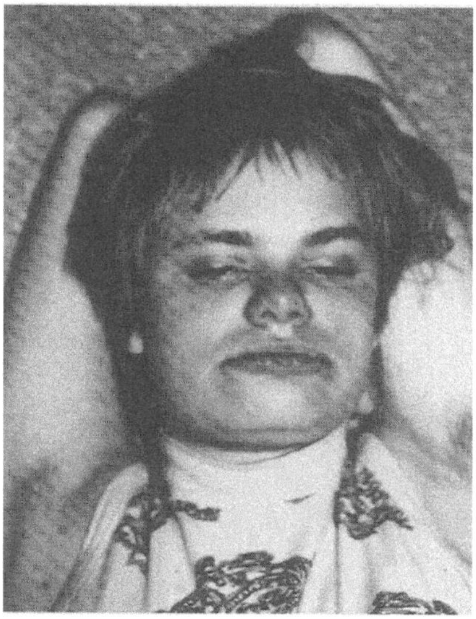

Figure 1. Case 1 at 20 years

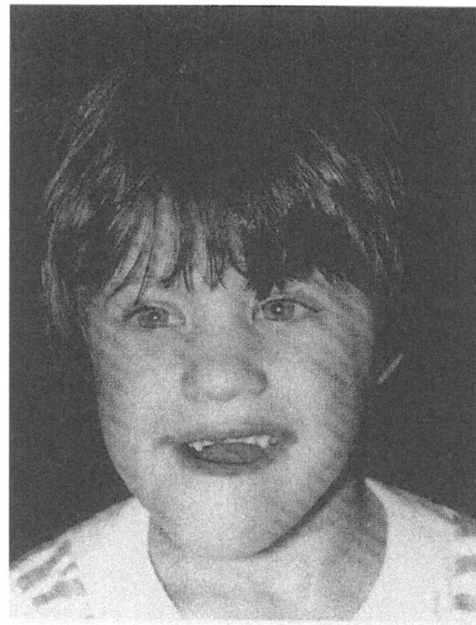

Figure 2. Case 2 at 5 1/2 years

Case 4: B.T. was a 3-year-old female. She had a normal perinatal history and a normal birth weight, height and head circumference. Delay was first noticed after the first half year of life. She first learned to walk at 2 years of age and never developed speech. She understood routine commands, some conversational speech, and recognized names of some objects. She had learned five hand signs. She did not eat well the first year of life. Her weight at 14 months was 7.9kg (<5th centile). She did not have seizures and an EEG report was normal. CT scan imaging showed mild atrophy. At 3 years, her height and weight were normal and her head circumference was <5th centile. Her hand and foot size were normal. She had hypotonia with increased deep tendon reflexes. Her gait was ataxic and jerky and she walked with outstretched arms. She had widely spaced teeth and exotropia. She had no hypopigmentation.

Case 5: M.P. was a 5-year-old female. She had a normal perinatal history. She had difficulty sleeping nights in the neonatal period and difficulty chewing. Developmental delay was first noted at about 3 months. She slept at most 6 hours per day, but this 6 hours was not continuous. She followed simple commands but had no speech. She had seizures at about 2 1/2 to 3 years but was now seizure free. Reports of an EEG and CT scan were normal. Her height was >97th centile and her weight was at the 90th centile. Her head circumference was at the 25th centile. Her hand and foot length were both <5th centile. Her tone was decreased but deep tendon reflexes were normal. She walked with an ataxic, jerky gait with outstretched arms. Her skin was hypopigmented. Of note, her parents' height and head size were also >97th centile.

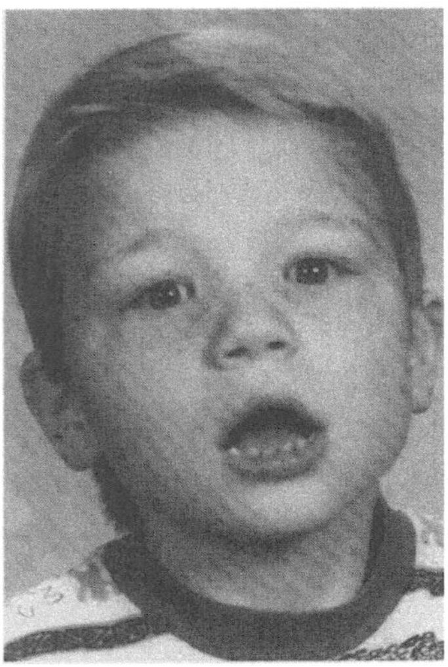

Figure 3. Case 3 at 6 years.

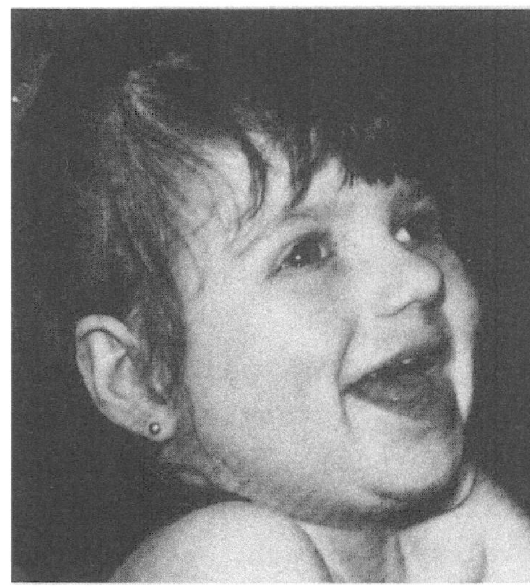

Figure 4. Case 4 at 3 years

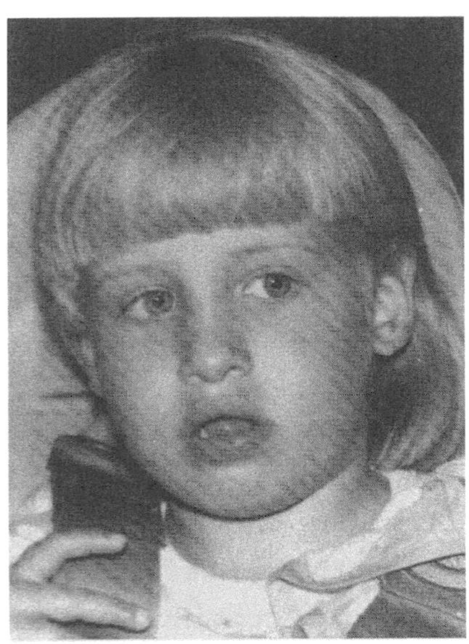

Figure 5. Case 5 at 5 years

DISCUSSION

Of the 26 individuals studied 5 did not have a demonstrable DNA deletion or uniparental disomy. Deletion negative individuals did not seem to differ clinically from deletion positive ones, given the clinical variability in AS. All had an unremarkable perinatal history, normal development in early infancy, physical exams consistent with AS and no other unrelated physical findings or medical problems. Case 4 and 5, however, were remarkable in that they were tall, and Case 4 had a head circumference at the 25th centile. Based on parental measurement, we feel that this is likely due to genetic background. It is interesting that none of the familial cases of AS has been reported to have a cytogenetic deletion (Kuroki et al, 1980; Pashayan et al, 1982; Fisher et al, 1987; Willems et al, 1987; Baraitser et al, 1987). Two sib pairs have been studied molecularly by Knoll et al (1991), and neither had a molecular deletion. One family with 3 affected siblings, however, has been recently studied molecularly and may have inherited a maternal deletion marked by probe 3-21 (see Niikawa et al, this Workshop).

The cause for nondeletional AS may be microdeletion, a recessive gene involving the same or other chromosomal regions, mosaicism, or an autosomal dominant gene modified by imprinting. Our 5 apparently sporadic cases could have the same cause or causes as the familial cases. Determining the genetic differences for nondeletional AS is an important counseling issue. If an autosomal recessive gene is involved, there would be up to a 25% recurrence risk, while involvement of an autosomal dominant gene would give up to a 50% recurrence risk.

REFERENCES

Baraitser M, Patton M, Lam STS, Brett EM and Wilson J (1987) The Angelman (happy puppet) syndrome: is it autosomal recessive? Clin Genet 31:323-330

Boyd SG, Harden A and Patton MA (1988) The EEG in early diagnosis of the Angelman (happy puppet) syndrome. Eur J Pediatr 147:508-513

Butler MG, Meaney FJ, Palmer CG (1986) Clinical and cytogenetic survey of 39 individuals with Prader-Labhart-Willi syndrome. Am J Med Genet 23:793-809

Butler MG and Palmer CG (1983) Parental origin of chromosome 15 deletion in Prader-Willi syndrome. Lancet 1:1285-1286

Fisher JA, Burn J, Alexander FW and Gardner-Medwin D (1987) Angelman (happy puppet) syndrome in a girl and her brother. J Med Genet 24:294-298

Knoll JHM, Glatt KA, Nicholls RD, Malcolm S, Lalande M (1991) Chromosome 15 uniparental disomy is not frequent in Angelman syndrome. Am J Hum Genet 48:16-21

Knoll JHM, Nicholls RD, Magenis RE, Glatt K, Graham JM Jr, Kaplan L and Lalande M (1990) Angelman syndrome: three molecular classes identified with chromosome 15q11q13-specific DNA markers. Am J Hum Genet 47:149-154

Knoll JHM, Nicholls RD, Magenis RE, Graham JM Jr, Lalande M, Latt SA (1989) Angelman and Prader-Willi syndromes share a common chromosome 15 deletion but differ in parental origin of the deletion. Am J Med Genet 32:285-290

Kuroki Y, Matsui I, Yamamoto Y and Ieshima A (1980) The "Happy Puppet" syndrome in two siblings. Hum Genet 56:227-229

Magenis RE, Brown MG, Lacy DA, Budden S, LaFranchi S (1987) Is Angelman syndrome an alternate result of del(15)(q11-q13)? Am J Med Genet 28:829-838

Malcolm S. Clayton-Smith J, Nichols M, Robb S. Webb T, Armour JAL, Jeffreys AJ and Pembrey ME (1991) Uniparental paternal disomy in Angelman's syndrome. Lancet 337:694-697

Nicholls RD, Knoll JHM, Butler MG, Karam S. Lalande M (1989a) Genomic imprinting suggested by maternal heterodisomy in non-deletion Prader-Willi syndrome. Nature 342:281-285

Nicholls RD, Knoll J. Glatt K, Hersh J, Brewster D, Graham J, Wurster-Hill D, Wharton R and Latt, S (1989b) Restriction fragment length polymorphisms within proximal 15q and their use in molecular cytogenetics and the Prader-Willi syndrome. Am J Med Genet 33:66-77

Pashayan H, Singer W, Bove C, Eisenberg E and Seto B (1982) The Angelman syndrome in two brothers. Am J Med Genet 13:295-298

Pembrey M, Fennell SJ, Van Den Berghe J, Fitchett M, Summers D, Butler L, Clarke C, Griffiths E, Thompson E, Super M and Baraitser M (1989) The association of Angelman's syndrome with deletions with 15q11-13. J Med Genet 26:73-77

Willems P, Dijkstra I, Brouwer O and Smit P (1987) Recurrence risk in the Angelman ("Happy Puppet") syndrome. Am J Med Genet 27:773-780

Williams CA, Zori RT, Stone JW, Gray BA, Hendrickson JE, Cantu ES, Ostrer H (1990) Maternal origin of 15q11-q13 deletions in Angelman syndrome suggests a role for genomic imprinting. Am J Med Genet 35:350-353

ACKNOWLEDGEMENT

This work was supported in part by funding from the Raymond C. Philips Research and Education Unit, the Department of Health and Rehabilitative Services, State of Florida. Editing and clerical assistance of Sandy Gentry is gratefully acknowledged.

GENETIC COUNSELING FOR ANGELMAN SYNDROME WHEN THE PROBAND HAS A CYTOGENETIC OR MOLECULAR DELETION

J. Hendrickson
Division of Genetics
Department of Pediatrics
Box 100296, University of Florida Health Science Center
Gainesville, Florida 32610-0296
USA

L. Marfatia[1], K. Kovak[2], R.E. Magenis[2], C. Williams[1]

University of Florida, Department of Pediatrics, Division of Genetics, Gainesville, Florida[1]; Oregon Health Sciences University, Child Development and Rehabilitation Center, Portland, Oregon[2], USA

INTRODUCTION

The Angelman Research and Support Group in the U.S.A. began at the University of Florida in the fall of 1987. Part of the impetus for formation of the organization was that families of the individuals with Angelman syndrome (AS) followed in our clinic both wanted to meet other families and had questions about the future of their child with AS which could not be adequately answered.

A major concern was, "How likely is this to happen again?"

METHODS

Willems et al. (1987) suggested a recurrence risk for AS of 1-2% after the birth of one child with AS, with the caveat that a 25% risk could not be ruled out. These data were based on previously published cases. Baraitser et al. (1987) reported three families with recurrence in siblings; in one case there were 3 affected siblings. He suggested that these cases provided further support that AS may be an autosomal recessive condition. Both papers were published prior to the recognition of the cytogenetic deletion in the 15q11-13 region.

NATO ASI Series, Vol. H 61
Prader-Willi Syndrome
Edited by S. B. Cassidy

To date, we are aware of only one case of recurrence in siblings when the proband has an identifiable deletion (Niikawa, this Workshop). Individuals with a deletion comprise an estimated 80 to 85% of all cases. The subsequently deleted chromosome 15 has been of maternal origin in all reported cases with the exception of a small number of paternal uniparental disomy cases in which no deletion has been identified. Included in our study are 23 deletion positive cases from a series of 29 probands examined and studied at the University of Florida; 82% had a deletion (Zachowski, in press).

To address the question of recurrence risk, our population of individuals with AS was divided into two groups: deletion-positive and deletion-negative. This report will address the statistical recurrence risk in siblings for the subgroup of deletion-positive individuals with AS.

The 46 probands with AS included in this study had diagnoses clinically confirmed by medical geneticists either at the University of Florida or at Oregon Health Sciences University. Detailed family history information was obtained by a genetic counselor. In all cases, the presence of a deletion was confirmed at the cytogenetic and/or molecular level.

RESULTS

Twenty-eight of the 46 deletion positive AS probands were studied at the University of Florida and 13 were primarily studied at Oregon Health Sciences University. An additional 2 families were examined by members at both institutions.

Age range of the 46 deletion positive probands was 2 years to 45 years. Ten had no siblings. The remaining 36 probands had a total of 61 full siblings, 9 maternal half siblings, and 6 paternal half siblings. None of the siblings had diagnoses or histories consistent with AS (Table I).

Table I. Pedigree Data Reported by 46 AS Families

Problems reported in siblings of the probands:

- 3 had seizures (none were major seizure disorders; all 3 had normal intelligence)
- 1 had learning disabilities
- 1 had an Attention Deficit Disorder
- 1 had craniosynostosis
- 1 maternal half-sib reportedly had Down syndrome

Problems reported in the probands' maternal family history (relationship given is to the mother):

- 1 sibling died at birth of a severe bone dysplasia, possibly osteogenesis imperfecta
- 1 uncle was reported to have seizures
- 1 uncle had motor delay
- 1 nephew had learning disabilities
- 2 cousins reportedly had isolated seizures
- 2 cousins had developmental delays
- 1 cousin each was reported to have autism, speech delay, and mental retardation with seizures
- 1 mother had adult polycystic kidney disease with obvious dominant family history of the disorder
- 2 families reported multiple members had depressive illness, with history of suicide in two members of one family

Problems reported in the probands' paternal family history (relationship given is to the father):

- 1 uncle had a speech impediment
- 1 cousin had seizures
- 1 cousin each had Down syndrome, Beckwith-Wiedemann syndrome, and Turner syndrome, and one family had multiple members with depressive illness

The data presented in Table I have not been evaluated for statistical significance in relationship to AS.

Two of the 46 probands had family histories which were remarkable.

Case 1 (see Pedigree 1) is that of a 5 year-old girl with classical AS findings of microbrachycephaly, ataxia, absent speech, frequent laughter, seizures, light pigmentation, and a cytogenetic deletion of 15q11-13. The mother reported that, as a child, she herself had experienced motor and language delays, walked at age 2, and had continued difficulty with coordination. She appeared to be moderately brachycephalic on examination and this was confirmed by computerized tomography (CT) scan. Molecular studies completed 3 years ago revealed that the chromosome 15 subsequently deleted in the child was maternally derived, but the mother had no apparent cytogenetic deletion (Williams et al, 1989). It may be that the mother's problems are coincidental, but it remains possible that her problems may be the result of some type of instability or change in the molecular region subsequently deleted in her child. Further molecular studies are planned.

Case 2 (see Pedigree 2) involves two independently ascertained relatives, second cousins with AS. The now 23 year-old female proband (IV-4) has been followed at the University of Florida with the diagnosis of AS since age 11. A deletion was demonstrated in 1989. She remains ataxic with absent speech and fewer episodes of excessive laughter. Her second cousin, a male now age 6, was initially diagnosed at another genetics center; both chromosomal and molecular studies revealed a deletion. The two families had lost contact with one another years ago but subsequently learned of one another through a directory published in the Angelman Support and Research Group newsletter. We have confirmed that the deleted chromosome 15 in the 6 year-old proband (IV-1) was maternally derived. Molecular studies on the young woman's family are in progress. It is noteworthy that the linking member of this family is the female proband's father, now unfortunately decreased.

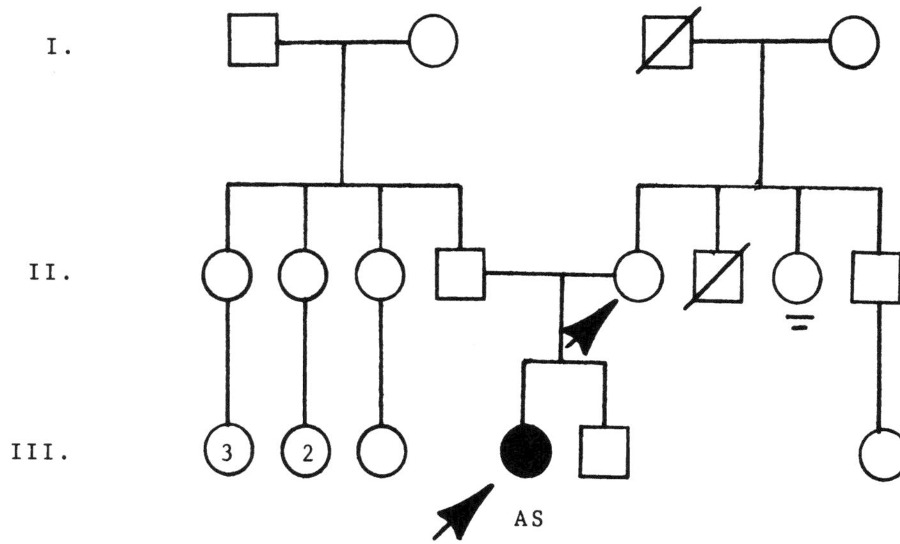

Pedigree 1: III-4 has a classical AS phenotype and is deletion positive; her mother, II-5, is deletion negative but has a history of speech and motor delays, ataxia, and brachycephaly.

DISCUSSION

If we consider counseling strictly for sibling recurrence, our data reveal no recurrences in 61 full siblings of 36 AS probands with deletions. Using exact confidence interval statistical methods for binomial data, we obtain a recurrence risk of less than 5% at the 95% confidence level (Shuster, 1988).

When providing recurrence risk counseling for AS, it is fortunate that a majority of probands have deletions, and thus markers. When counseling parents of deletion positive AS individuals, we quote less than a 5% risk, qualifying the information by stating 1) with one exception, we are aware of no cases of recurrence in siblings when the proband is deletion positive. Niikawa (this Workshop) reported a family in which three AS siblings as well as their normal mother and maternal grandfather have molecular deletions, and 2) the statistical method used to determine this risk estimate is very conservative, requiring a large data pool to further reduce the risk. Prenatal diagnosis by karyotyping and molecular analysis are discussed as possible options.

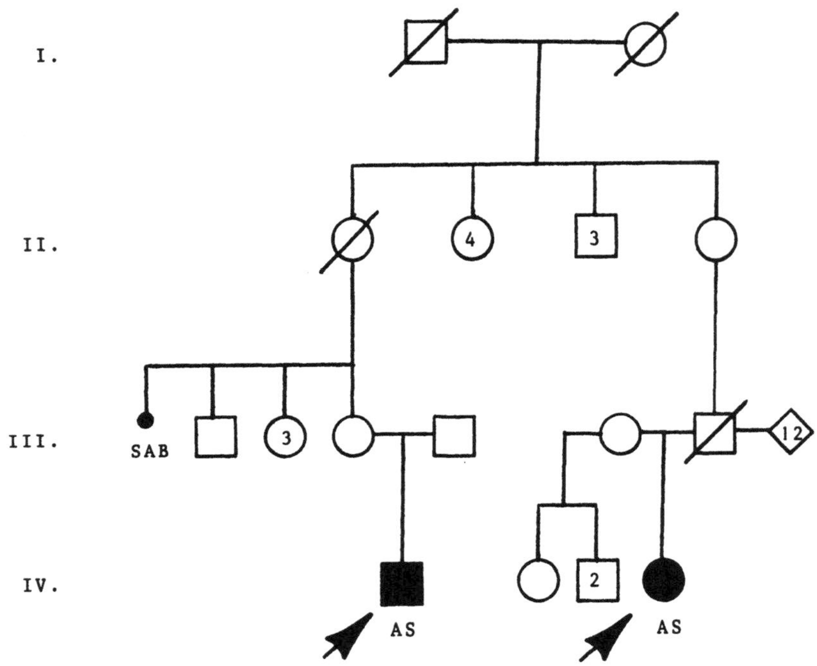

Pedigree 2: IV-1 and IV-4 have classical AS phenotypes and are both deletion positive. Note that III-6 is a linking male.

The kindred with deletion positive second cousins and the family in which the mother may have mild phenotypic expression suggest that some unique deletion families may have increased risks which cannot be quantified at this time. These families are very important since they may provide additional insight into our understanding of this molecular region.

Recurrence in siblings has been reported in families who are deletion negative by current molecular techniques (Knoll, 1991; Malcolm, 1991). Only 5 families thoroughly studied at the University of Florida remain deletion negative. These families are small; 2 have no siblings, 3 have one sibling, and one has

2 siblings, so it is not possible to adequately address recurrence risk. When the proband is deletion negative, the risk to siblings is probably variable, and probably higher than 5%, although it cannot be presently quantified. Without a marker, prenatal diagnosis is not possible.

An important goal is to provide more accurate and more complete recurrence risk information for families having a child with AS. This may be best accomplished by continuing to build a data pool of AS families through clinical efforts, the Angelman Support and Research Group, and through collaboration with other interested institutions.

REFERENCES

Baraitser M, Patton M, Lam STS, Brett EM, Wilson J (1987) The Angelman (happy puppet) syndrome: Is it autosomal recessive? Clin Genet 31:323-330

Knoll JHM, Glatt K, Nicholls RD, Malcolm S, Lalande M (1991) Chromosome 15 uniparental disomy is not frequent in Angelman syndrome. Am J Hum Genet 48:16-21

Malcolm S, Nichols M, Clayton-Smith J, Robb S, Webb T, Armour JAL, Jeffreys AG, Pembrey ME (1991) Uniparental paternal disomy in Angelman syndrome. Lancet 337:696-697

Shuster JJ (1988) EXACTB CONFL: Exact unconditional procedures for binomial data. Am Statistician 42:234

Willems PJ, Dijkstra I, Oebele F, Smit PG (1987) Recurrence risk in the Angelman (happy puppet) syndrome. Am J Med Genet 27:773-780

Williams CA, Hendrickson JE, Cantu ES, Donlon TA (1989) Angelman syndrome in a daughter with del(15)(q11q13) associated with brachycephaly, hearing loss, enlarged foramen magnum, and ataxia in the mother. Am J Med Genet 32(3):333-338

Zackowski JZ, Nicholls RD, Gray BA, Bent-Williams A, Gottlieb W, Harris PJ, Waters MF, Driscoll DJ, Zori RT, Williams CA (in press) Cytogenetic and molecular analysis in Angelman syndrome. Am J Med Genet

CURRENT UNDERSTANDING AND RECURRENCE RISKS OF PRADER-WILLI AND ANGELMAN SYNDROMES

Timothy A. Donlon, Ph.D.
Department of Pathology
Stanford University Medical Center
300 Pasteur Drive
Stanford, California 94305
USA

Panel Discussion by: Bernhard Horsthemke, Ph.D., Institut für Humangenetik, Universitätsklinikum Essen, D-4500 Essen, Germany; Ellen Magenis, M.D., Child Development and Rehabilitation Center, Oregon Health Sciences Center, 707 S.W. Gaines Road, Portland, Oregon, 97201, USA; Marcus Pembrey, M.D., Institute for Child Health, University of London, 30 Guilford Street, London WC1N 1EH, England, and lead by Timothy A. Donlon, Ph.D.

The goal of this report is to present a brief summary of the Workshop and our current understanding regarding the laboratory diagnosis (primarily molecular) of microdeletions associated with Prader-Willi syndrome (PWS) and Angelman syndrome (AS). Since much of this information is anecdotal, it is intended more as a report of the general level of progress that has been made in the analyses of 15q microdeletion syndromes rather than of a complete chronicle of those who have contributed to this prodigious topic.

DNA ANALYSIS

Most posters and platform presentations at this Workshop regarding the DNA analysis of 15q deletions have focused on DNA segments cloned by two groups, one involving probes from a flow-sorted chromosome library (Donlon, 1986) and the other from a microdissected library (Horsthemke, 1989). Since the latter probes have thus far been largely uninformative for

NATO ASI Series, Vol. H 61
Prader-Willi Syndrome
Edited by S. B. Cassidy

polymorphisms, most analyses have focused on probes from the flow-sorted library. Unfortunately, most of these analyses have included only five loci that are commonly deleted as a unit, so that it is not possible to determine the level of heterogeneity in deletion breakpoints between patients (see Continguous Gene Syndromes on 15q, below).

The data presented at this Workshop fall into the two general categories of frequencies and parental origin of deletions. There is general agreement that a good correlation exists between cytogenetically and molecularly detected deletions in PWS and AS. Individual reports have surfaced presenting cases of molecular deletion in patients with normal chromosomes, but some of these cases appear to provide less resolution than necessary to rule out a subtle cytogenetic deletion. There are also cases with apparent cytologic deletion but with no molecular deletion. It is not yet possible to conclude whether these might be deletions of other contiguous DNA segments (i.e., other than the five loci that most investigators have been examining) or might result from "cryptic translocations", which are translocations that are either too small to be detected microscopically or involve chromosomal regions which stain similarly.

The frequency of 15q deletion in PWS patients is approaching 77% based upon a combination of molecular and cytogenetic analyses. Most molecular analyses have focused on the use of restriction fragment length polymorphisms (RFLPs), and some groups have increased the number of patients that may be analyzed by using the less reliable method of dosage blot hybridization since some families are not informative for the RFLPs. Pembry and coworkers presented data at this Workshop on 84 AS patients from 78 families; twelve patients were, therefore, familial. Of the 66 families that had a nonfamilial case of AS, 46 (70%) showed de novo 15q deletion. Nicholls and coworkers then reported deletions in 19 of 25 patients (76%) using a combination of RFLP and dosage blot analysis.

Uniparental disomy (UPD) is a state in which an individual receives two members of one pair of chromosomes from one parent and none from the other. In order for analysis for disomy to be accurate, a polymorphism of three or more alleles must be employed, otherwise one must use the less reliable methodology of dosage blot hybridization to distinguish hemi- from homozygosity. Malcolm and coworkers (this Workshop) reported the use of a probe for a multi-allele locus (D15S86) near the telomere of 15q to examine this issue in PWS and AS. They found UPD to be more prevalent in PWS; roughtly 23% in PWS vs. 4% in AS. These conclusions have been supported by Nicholls and coworkers (this Workshop). Table I compares and contrasts the frequencies of deletions and UPD in PWS and AS. Based on these data, it appears that the etiology for all or most cases of PWS fit neatly into two groups, deletion or uniparental disomy. On the other hand, only 80% of the AS cases have a known molecular abnormality (76% for deletion and 4% for UPD) leaving many cases (20%) unexplained (see Contiguous Gene Syndromes, below).

PARENTAL ORIGIN AND GENOMIC IMPRINTING

The maternal origin of *de novo* 15q11.2 deletions in AS and the paternal origin of similar deletions in PWS strongly implicates the phenomenon of "genomic imprinting" in the etiology of these two disparate syndromes. This term is used to denote modification of a portion of the genome (possibly inactivation) by the passage of that genome through a meiotic cycle (Surani, 1984). While little is known regarding the mechanism of imprinting, the data from mouse experiments imply three types of imprinting: (1) those chromosomal regions that are imprinted in male meiosis, (2) those that are imprinted in female meiosis, and (3) those regions that are differentially imprinted in both sexes. Thus, normal development requires genetic contribution from the mother, father, or from both parents, in the three respective types. Human chromosome 15 band q11.2 is a region that appears to be under the influence of the third type, or differential imprinting. The Prader-Willi syndrome appears to be attributable to a failure to inherit a gene or set of genes from the father, either by deletion or by UPD, whereas Angelman syndrome appears to result from a failure to inherit a nearby but not necessarily identical maternal locus or loci.

Why are the frequencies of UPD in PWS and AS different? If UPD generally arises through a trisomy 15 intermediate followed by the random loss of one homolog, then an increased rate of nondisjunction in older mothers might help to explain the higher incidence of UPD in PWS. This does not explain the 20% of AS patients who have neither a deletion nor UPD, but these may be attributable to deletions that are not readily detected using the standard five probes. In another comparison, why has PWS and not AS been observed in patients who have an extra, inverted duplicated chromosome 15 (inv dup(15)) without deletion? These cases presumably have four copies of the same region in which other patients are hemizygous and yet some of them have typical PWS. The answer to this question may be related to the presentation by Surani (this Workshop) suggesting that an exact ratio of maternal to paternal alleles is necessary for normal development. Since most (about 85%) of these extra inv dup(15)s are maternally derived, the affected patients have a 3:1 ratio of maternal to paternal alleles, which may be functionally similar to having a 1:0 ratio.

Table I

APPROXIMATE FREQUENCIES OF DETECTABLE DELETIONS

	PWS	AS
de novo Deletion	77%	76%
Uniparental Disomy	23%	4%
Other	0	20%

Table II

ESTIMATED RECURRENCE RISKS

	PWS	AS
de novo Deletion	<1%	<2%
UPD, Normal Parents	<1%	2%
Sporadic, Non-Deletion, UPD excluded	<1%	18-25%?
Familial, Non-Deletion, UPD excluded	<1%	50%?
UPD with Familial Translocation	?	5-10%?

RECURRENCE RISK

There have been very few cases of familial PWS reported in the literature, and those cases have been cytogenetically normal. Cassidy (1987) has reported this frequency to be less than 1% based on the total number of families in the American PWS registry. Pembrey et al (this Workshop) presented data on 84 AS patients from 78 families some of which had multiple affected sibs. Their findings suggest that in those cases in which the proband has a _de novo_ deletion the recurrence risk is low. If parents are cytogenetically normal the recurrence risk for affected children with uniparental disomy is low in both PWS and AS. However, there is reason to believe that if one parent is a carrier for a chromosome 15 rearrangement then the recurrence risk for UPD or deletion may be elevated due to nondisjunction or aberrant recombination. Their data suggest that this risk may be as high as 5-10%, although more data are needed to clarify this latter issue. If the affected individual has UPD and the parents are cytogenetically normal, then the recurrence risk for UPD may be elevated due to the role that nondisjunction has been suggested to play in the production of UPD. This risk may be approximately 2%, although again more data are needed. If there is a horizontal distribution of AS in a family, the parents are cytogenetically normal, and UPD has been excluded, then one cannot rule out an autosomal recessive mode of inheritance or a submicroscopic deletion. These proposed recurrence risks are summarized in Table II.

Some familial cases may behave more like autosomal dominant disorders due to the presence of affected patients in multiple branches of the fmaily. A family was presented by Niikawa and Hamabe (this Workshop) in which the mother of three affected sibs and her father all posses a submicroscopic deletion (locus D15S10 identified by probe pTD3-21). The maternal grandfather is unaffected presumably because he inherited the deletion from his father. A similar family is exemplified in Figure 1 in which a woman received a mutation from her unaffected father. Here, each of her children would be at 50% risk for inheriting the deletion and developing AS. Her carrier brother, who is also unaffected, would have a 50% chance of passing the mutation on to his offspring, although none of them would be affected with AS because this locus is presumably inactivated during male meiosis.

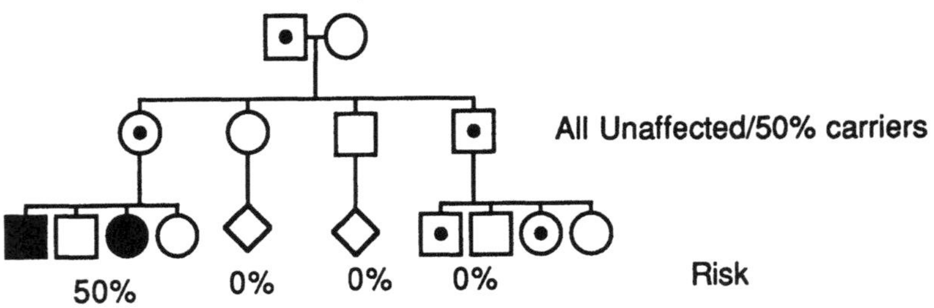

Figure 1. Presumed pattern of inheritance of familial AS

CONTIGUOUS GENE SYNDROMES ON 15q

A microdeletion syndrome, or contiguous gene syndrome, is believed to result from the loss or gain of contiguous genetic segments. The phenotypes will generally include a core pattern of abnormalities, or syndrome, resulting from the loss or gain of a common set of genes (critical region), yet there also may be a high degree of clinical variability that is presumed to reflect variation in the exact breakpoints around this critical region. This matter can be resolved only by molecular analysis since cytogenetic analysis is limited to about 3-4 million base pairs (Mb). The literature is replete with reports of patients who have suspected cytologic deletion but who have neither PWS nor AS. None of these non-PWS, non-AS patients have been confirmed using molecular methods unless they express some clinical overlap with either of the two syndromes. The conclusions from these data suggest either that there is no deletion or that the appropriate DNA probes have not been used for analysis. Unfortunately, most groups are examining the same five or six loci, identified by probes pIR4-3 (D15S11), pTD189-1 (D15S13), pML34 (D15S9), pTD3-21 (D15S10), pIR10-1 (D15S12), and sometimes pIR39 (D15S19). The consensus that is developing supports data from Knoll et al (1990) who examined eight loci in 19 AS patients. They indicated that there are generally three different groups of patients: (1) those who have no apparent deletion, (2) those who are missing the first five loci shown above, and (3) those who are missing pIR39 in addition to the previous five. The problem with this type of analysis is that these loci span a limited region of the presumed deletion. For example, pIR4-3, pTD189-1, and pML34 identify the same 0.7 Mb NotI fragment, while pTD3-21 and pIR10-1 identify the same 1.3 Mb SalI fragment (Donlon, 1991), so that these loci occupy no greater than 2 Mb of combined DNA and perhaps much less (see Figure 2). The sizes of these deletions have been estimated to be 3-4 Mb in size. At present, there is a gap of 1-2 Mb between these two sets of loci. So that it is not possible to discern the sizes and breakpoints for these deletions, nor to define critical regions responsible for PWS and AS. This will require the long-range mapping and deletion analysis of many loci along the entire 15q11-q13 region not only in typical PWS and AS patients but also in those patients who do not strictly fit the classical clinical definitions.

The finding that most, or all, PWS patients have either deletion of the five loci or UPD suggests that the loci that lead to PWS are bracketed by these five cloned DNA segments. In AS, however, 20% of the cases appear to be caused by a mechanism other than the loss of these five RFLPs and may well represent slightly different and smaller deletions. The data by Niikawa and coworkers (this Workshop) show that of these five probes pTD3-21 may be solely deleted in patients with AS, suggesting that the AS critical region is nested within that for PWS. The breakpoints for these small AS deletions could presumably lie within this gap between pML34 and pTD3-21.

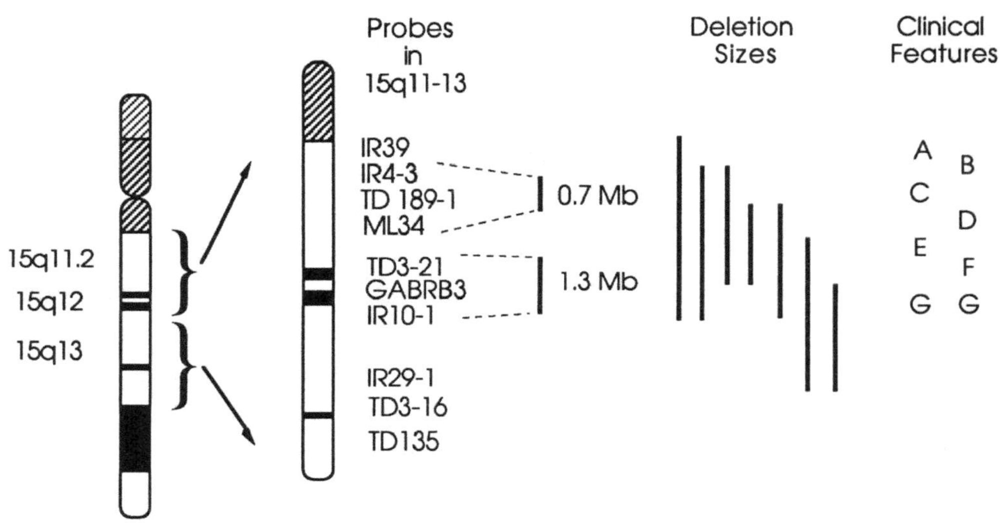

Figure 2. Distribution of DNA Markers in Chromosome 15q11.2-q12. (For explanation, please see text.)

The theory of contiguous gene syndromes states that the phenotype is determined by the size and location of the deleted segments. The phenotypes of PWS and AS also vary by the parent of origin in the deletion. The following is an hypothesis of how 15q deletion disorders may vary by the size, location, and inheritance of the deletions; this is illustrated in Figure 2.

Since PWS and AS are contiguous gene syndromes, they may be considered to be composites of genetic disorders such that the core of disorders, or complete syndrome, is caused by deletion of a core set of genes, the "critical region:. Genes for the clinical parameters of PWS (e.g., "lanuage disorders" - "A", truncal obesity - "C", hypotonicity - "E", and hypopigmentation - "G") could be caused by deletion of loci A-G. Similarly, genes involved in the expression of AS (e.g., absent speech - "B", poor motor coordination - "D", hypertonicity - "F", and hypopigmentation - "G") could include the same region of the chromosome, A-G. If differential imprinting of loci A-F occurs such that A, C, and E are inactivated in female-derived gametes and B. D. and F are inactivated in male-derived gametes, then a deletion of the paternal allele will eliminate the only set of functional copies of loci A, C, and E, while with a maternal deletion, functional copies of loci B, D, and F are eliminated. Loci that are not imprinted will be expressed on both parents' chromosomes and may lead to a characteristic phenotype regardless of the parental origin, but will rather depend on the size of the deletion. A feature such as hypopigmentation, which is found in both PWS and AS, would satisfy this latter type of locus, denoted by "G".

PREDICTING PHENOTYPES

A pediction of a person's phenotype is not possible at the current stage of laboratory diagnosis. If a clinician receives the results that his/her patient has a deletion of 15q11.2-q12, this information should be used only as confirmation of a suspected diagnosis of PWS or AS, keeping in mind that $\geq$23% of the clinically typical PWS patients have no recognizable deletion. As more detailed molecular analyses are avilable it should be possible to discern critical regions for each of these disorders. If these analyses conclude that there is no variation in exact breakpoints among patients with the same syndrome, then most or all clinical findings will always be found together, making phenotypic predictions somewhat straightforward. The range of clinical features for each disorder will thus be small. On the contrary, if there is variation in breakpoints, a situation that seems more likely, then a full prediction of the patient's phenotype will require a detailed analysis of perhaps four Mb of DNA (the estimated size of the cytologic deletion) in addition to the parental origin of this deleted material.

Another level of complexity in the prediction of phenotypes is found in those patients who do not have a deletion but instead have uniparental disomy for chromosome 15. The exact type of UPD (i.e., iso- vs. heterodisomy) requires informative markers along the entire length of the chromosome, since meiotic exchange can produce a chromosome that is isodisomic at the centromeres and heterodisomic at more distal markers (a second meiotic error). The reverse situation may exist for disomic chromosomes produced by first meiotic errors. Some PWS or AS patients will be found to have additional genetic abnormalities that are unrelated to either PWS or AS due to UPD. These additional abnormalities may result from UPD because there will be chromosomal segments that are isodisomic (homozygous) for abnormal genes that map well outside the PWS and AS critical regions.

STANDARDS FOR DESCRIBING PHENOTYPES

The detailed assignment of chromosomal regions that are critical for the expression of genetic syndromes will require not only additional cloned DNA segments but also a detailed description of well defined anthropomorphic, neurologic, and metabolic parameters. To address issues relating to recurrence risks, diagnosis, and prognosis it will be necessary to examine classical cases of PWS and AS as well as those cases which deviate from the more rigorous definitions. Only by including these diverse cases can we determine a relationship between well defined clinical syndromes with common regions of deletion and other types of chromosome abnormality. Several presentations at this Workshop involved detailed clinical descriptions of the PWS and AS phenotypes as well as comparisons between different ethnic groups which should help to guide clinicians in this attempt.

REFERENCES

Cassidy SB (1987) Recurrence risk in Prader-Willi syndrome. Am J Med Genet 28:59-60

Donlon TA (1991) unpublished data

Donlon TA, Lalande M, Wyman A, Bruns G, Latt SA (1986) Isolation of molecular probes associated with the chromosome 15 instability in the Prader-Willi syndrome. Proc Nat Acad Sci 83:4408-4412

Horsthemke B, Buiting K, Lüdecke H-J, Neumann M, Senger G, Claussen U, Grzeschik K-H, Passarge E (1989) New probes for the Prader-Willi syndrome chromosome region (PWCR) and other loci on chromosome 15. Cytogenet Cell Genet 51:1015

Knoll JHM, Nicholls RD, Magenis RE, Glatt K, Graham JM Jr, Kaplan L, Lalande M (1990) Angelman syndrome: Three molecular classes identified with chromosome 15q11q13-specific DNA markers. Am J Hum Genet 47:149-155

Surani MAH, Barton SC, Norris LM (1984) Development of reconstituted mouse eggs suggests imprinting of the genome during gametogenesis. Nature 308:548-550

NATO ASI Series H

Vol. 1: **Biology and Molecular Biology of Plant-Pathogen Interactions.** Edited by J. A. Bailey. 415 pages. 1986.

Vol. 2: **Glial-Neuronal Communication in Development and Regeneration.** Edited by H. H. Althaus and W. Seifert. 865 pages. 1987.

Vol. 3: **Nicotinic Acetylcholine Receptor: Structure and Function.** Edited by A. Maelicke. 489 pages. 1986.

Vol. 4: **Recognition in Microbe-Plant Symbiotic and Pathogenic Interactions.** Edited by B. Lugtenberg. 449 pages. 1986.

Vol. 5: **Mesenchymal-Epithelial Interactions in Neural Development.** Edited by J. R. Wolff, J. Sievers, and M. Berry. 428 pages. 1987.

Vol. 6: **Molecular Mechanisms of Desensitization to Signal Molecules.** Edited by T. M. Konijn, P. J. M. Van Haastert, H. Van der Starre, H. Van der Wel, and M. D. Houslay. 336 pages. 1987.

Vol. 7: **Gangliosides and Modulation of Neuronal Functions.** Edited by H. Rahmann. 647 pages. 1987.

Vol. 8: **Molecular and Cellular Aspects of Erythropoietin and Erythropoiesis.** Edited by I. N. Rich. 460 pages. 1987.

Vol. 9: **Modification of Cell to Cell Signals During Normal and Pathological Aging.** Edited by S. Govoni and F. Battaini. 297 pages. 1987.

Vol. 10: **Plant Hormone Receptors.** Edited by D. Klämbt. 319 pages. 1987.

Vol. 11: **Host-Parasite Cellular and Molecular Interactions in Protozoal Infections.** Edited by K.-P. Chang and D. Snary. 425 pages. 1987.

Vol. 12: **The Cell Surface in Signal Transduction.** Edited by E. Wagner, H. Greppin, and B. Millet. 243 pages. 1987.

Vol. 13: **Toxicology of Pesticides: Experimental, Clinical and Regulatory Perspectives.** Edited by L. G. Costa, C. L. Galli, and S. D. Murphy. 320 pages. 1987.

Vol. 14: **Genetics of Translation. New Approaches.** Edited by M. F. Tuite, M. Picard, and M. Bolotin-Fukuhara. 524 pages. 1988.

Vol. 15: **Photosensitisation. Molecular, Cellular and Medical Aspects.** Edited by G. Moreno, R. H. Pottier, and T. G. Truscott. 521 pages. 1988.

Vol. 16: **Membrane Biogenesis.** Edited by J. A. F. Op den Kamp. 477 pages. 1988.

Vol. 17: **Cell to Cell Signals in Plant, Animal and Microbial Symbiosis.** Edited by S. Scannerini, D. Smith, P. Bonfante-Fasolo, and V. Gianinazzi-Pearson. 414 pages. 1988.

Vol. 18: **Plant Cell Biotechnology.** Edited by M. S. S. Pais, F. Mavituna, and J. M. Novais. 500 pages. 1988.

Vol. 19: **Modulation of Synaptic Transmission and Plasticity in Nervous Systems.** Edited by G. Hertting and H.-C. Spatz. 457 pages. 1988.

Vol. 20: **Amino Acid Availability and Brain Function in Health and Disease.** Edited by G. Huether. 487 pages. 1988.

NATO ASI Series H

Vol. 21: **Cellular and Molecular Basis of Synaptic Transmission.** Edited by H. Zimmermann. 547 pages. 1988.

Vol. 22: **Neural Development and Regeneration. Cellular and Molecular Aspects.** Edited by A. Gorio, J. R. Perez-Polo, J. de Vellis, and B. Haber. 711 pages. 1988.

Vol. 23: **The Semiotics of Cellular Communication in the Immune System.** Edited by E. E. Sercarz, F. Celada, N. A. Mitchison, and T. Tada. 326 pages. 1988.

Vol. 24: **Bacteria, Complement and the Phagocytic Cell.** Edited by F. C. Cabello und C. Pruzzo. 372 pages. 1988.

Vol. 25: **Nicotinic Acetylcholine Receptors in the Nervous System.** Edited by F. Clementi, C. Gotti, and E. Sher. 424 pages. 1988.

Vol. 26: **Cell to Cell Signals in Mammalian Development.** Edited by S. W. de Laat, J. G. Bluemink, and C. L. Mummery. 322 pages. 1989.

Vol. 27: **Phytotoxins and Plant Pathogenesis.** Edited by A. Graniti, R. D. Durbin, and A. Ballio. 508 pages. 1989.

Vol. 28: **Vascular Wilt Diseases of Plants. Basic Studies and Control.** Edited by E. C. Tjamos and C. H. Beckman. 590 pages. 1989.

Vol. 29: **Receptors, Membrane Transport and Signal Transduction.** Edited by A. E. Evangelopoulos, J. P. Changeux, L. Packer, T. G. Sotiroudis, and K. W. A. Wirtz. 387 pages. 1989.

Vol. 30: **Effects of Mineral Dusts on Cells.** Edited by B. T. Mossman and R. O. Bégin. 470 pages. 1989.

Vol. 31: **Neurobiology of the Inner Retina.** Edited by R. Weiler and N. N. Osborne. 529 pages. 1989.

Vol. 32: **Molecular Biology of Neuroreceptors and Ion Channels.** Edited by A. Maelicke. 675 pages. 1989.

Vol. 33: **Regulatory Mechanisms of Neuron to Vessel Communication in Brain.** Edited by F. Battaini, S. Govoni, M. S. Magnoni, and M. Trabucchi. 416 pages. 1989.

Vol. 34: **Vectors as Tools for the Study of Normal and Abnormal Growth and Differentiation.** Edited by H. Lother, R. Dernick, and W. Ostertag. 477 pages. 1989.

Vol. 35: **Cell Separation in Plants: Physiology, Biochemistry and Molecular Biology.** Edited by D. J. Osborne and M. B. Jackson. 449 pages. 1989.

Vol. 36: **Signal Molecules in Plants and Plant-Microbe Interactions.** Edited by B. J. J. Lugtenberg. 425 pages. 1989.

Vol. 37: **Tin-Based Antitumour Drugs.** Edited by M. Gielen. 226 pages. 1990.

Vol. 38: **The Molecular Biology of Autoimmune Disease.** Edited by A. G. Demaine, J-P. Banga, and A. M. McGregor. 404 pages. 1990.

Vol. 39: **Chemosensory Information Processing.** Edited by D. Schild. 403 pages. 1990.

Vol. 40: **Dynamics and Biogenesis of Membranes.** Edited by J. A. F. Op den Kamp. 367 pages. 1990.

Vol. 41: **Recognition and Response in Plant-Virus Interactions.** Edited by R. S. S. Fraser. 467 pages. 1990.

NATO ASI Series H

Vol. 42: **Biomechanics of Active Movement and Deformation of Cells.**
Edited by N. Akkaş. 524 pages. 1990.

Vol. 43: **Cellular and Molecular Biology of Myelination.**
Edited by G. Jeserich, H. H. Althaus, and T. V. Waehneldt. 565 pages. 1990.

Vol. 44: **Activation and Desensitization of Transducing Pathways.**
Edited by T. M. Konijn, M. D. Houslay, and P. J. M. Van Haastert. 336 pages. 1990.

Vol. 45: **Mechanism of Fertilization: Plants to Humans.**
Edited by B. Dale. 710 pages. 1990.

Vol. 46: **Parallels in Cell to Cell Junctions in Plants and Animals.**
Edited by A. W. Robards, W. J. Lucas, J. D. Pitts, H. J. Jongsma, and D. C. Spray. 296 pages. 1990.

Vol. 47: **Signal Perception and Transduction in Higher Plants.**
Edited by R. Ranjeva and A. M. Boudet. 357 pages. 1990.

Vol. 48: **Calcium Transport and Intracellular Calcium Homeostasis.**
Edited by D. Pansu and F. Bronner. 456 pages. 1990.

Vol. 49: **Post-Transcriptional Control of Gene Expression.**
Edited by J. E. G. McCarthy and M. F. Tuite. 671 pages. 1990.

Vol. 50: **Phytochrome Properties and Biological Action.**
Edited by B. Thomas and C. B. Johnson. 337 pages. 1991.

Vol. 51: **Cell to Cell Signals in Plants and Animals.**
Edited by V. Neuhoff and J. Friend. 404 pages. 1991.

Vol. 52: **Biological Signal Transduction.**
Edited by E. M. Ross and K. W. A. Wirtz. 560 pages. 1991.

Vol. 53: **Fungal Cell Wall and Immune Response.**
Edited by J. P. Latgé and D. Boucias. 472 pages. 1991.

Vol. 54: **The Early Effects of Radiation on DNA.**
Edited by E. M. Fielden and P. O'Neill. 448 pages. 1991.

Vol. 55: **The Translational Apparatus of Photosynthetic Organelles.**
Edited by R. Mache, E. Stutz, and A. R. Subramanian. 260 pages. 1991.

Vol. 56: **Cellular Regulation by Protein Phosphorylation.**
Edited by L. M. G. Heilmeyer, Jr. 520 pages. 1991.

Vol. 57: **Molecular Techniques in Taxonomy.**
Edited by G. M. Hewitt, A. W. B. Johnston, and J. P. W. Young. 420 pages. 1991.

Vol. 58: **Neurocytochemical Methods.**
Edited by A. Calas and D. Eugène. 352 pages. 1991.

Vol. 59: **Molecular Evolution of the Major Histocompatibility Complex.**
Edited by J. Klein and D. Klein. 522 pages. 1991

Vol. 60: **Intracellular Regulation of Ion Channels.**
Edited by M. Morad and Z. Agus. 261 pages. 1992.

Vol. 61: **Prader-Willi Syndrome and Other Chromosome 15 q Deletion Disorders.**
Edited by S. B. Cassidy. 281 pages. 1992

MIX
Papier aus verantwortungsvollen Quellen
Paper from responsible sources
FSC® C105338

If you have any concerns about our products,
you can contact us on
ProductSafety@springernature.com

In case Publisher is established outside the EU,
the EU authorized representative is:
Springer Nature Customer Service Center GmbH
Europaplatz 3, 69115 Heidelberg, Germany

Printed by Libri Plureos GmbH
in Hamburg, Germany